Fast Facts for Healthcare Professionals

Gastroenterology

The Gut Microbiome

Fergus Shanahan MD DSc
Professor emeritus of Medicine
University College Cork
National University of Ireland and
APC Microbiome Ireland
Cork, Ireland

Declaration of Independence
This book is as balanced and as practical as we can make it.
Ideas for improvement are always welcome: fastfacts@karger.com

HEALTHCARE

Fast Facts: The Gut Microbiome
First published 2024

Text © 2024 Fergus Shanahan
© 2024 in this edition S. Karger Publishers Limited

S. Karger Publishers Limited, Merchant House, 5 East St. Helen Street, Abingdon,
Oxford OX14 5EG, UK

Book orders can be placed by telephone or email, or via the website.
Please telephone +41 61 306 1440 or email orders@karger.com
To order via the website, please go to karger.com

Fast Facts is a trademark of S. Karger Publishers Limited.

All rights reserved. No part of this publication may be reproduced, stored in a retrieval system, or transmitted in any form or by any means, electronic, mechanical, photocopying, recording, or otherwise, without the express permission of the publisher.

The rights of Fergus Shanahan to be identified as the authors of this work have been asserted in accordance with the Copyright, Designs & Patents Act 1988 Sections 77 and 78.

The publisher and the authors have made every effort to ensure the accuracy of this book, but cannot accept responsibility for any errors or omissions. For all drugs, please consult the product labeling approved in your country for prescribing information.

Registered names, trademarks, etc. used in this book, even when not marked as such, are not to be considered unprotected by law.

A CIP record for this title is available from the British Library.

ISBN: 978-3-318-07292-1

Shanahan F (Fergus)
Fast Facts: The Gut Microbiome/
Fergus Shanahan

Medical illustrations by Graeme Chambers, Belfast, UK.
Typesetting by Amnet, Chennai, India.
Printed and bound by CPI Group (UK) Ltd, Croydon, CR0 4YY

Foreword	5
Glossary of terms	7
List of abbreviations	9
Introduction	11
Life without microbes	23
What's normal?	31
The changing microbiome as the host ages	41
Food and microbes	51
Mindful microbes: brain–gut signaling	67
Microbes and chronic intestinal disease	75
Cancer and the microbiome	87
Drugs and the microbiota	97
Antimicrobial resistance	107
The gut virome and mycobiome	117
Therapeutic modification of the microbiome	129
Index	139

Foreword

Whether you are a scientist, a clinician or a member of the general population, you will be more than aware of the deluge of information that assaults us in relation to the microbes that co-exist with us – our microbiomes. Whether you source your information from the most prestigious journals in biomedical science or surf the web, you are likely to be overwhelmed by the volume of new knowledge on this topic and may well struggle to make sense of it all.

The microbiome that inhabits the gastrointestinal tract has, rightly or wrongly, attracted the greatest attention and generated the most research. Skimming through the medical literature on the gut microbiome one could be forgiven for concluding that gut microbes are the source of all our ills and that every effort should be directed towards the development of interventions that restore microbiome homeostasis.

A critical review of the status of the gut microbiome will, therefore, be greeted with great enthusiasm. This is no simple task and one that few are as qualified as Fergus Shanahan to achieve. A gastroenterologist with an international reputation in the area of inflammatory bowel disease (a vivid illustration of the complexities of microbe-host interactions), he is equally at home in the science of the microbiome and, as founding director of the Alimentary Pharmabiotic Centre in Cork, Ireland, led a multidisciplinary team of clinicians and scientists who were at the very forefront of microbiome science.

The depth of Professor Shanahan's knowledge of the science of the microbiome is evident throughout but what transforms this volume from a mere recitation of what we know to a truly critical and accessible exposition on what matters is the critical eye of the clinician who must make decisions that will affect a patient's health. The tone of this wonderfully written and illustrated volume is set at the outset – 'this book is about the known facts'.

What follows, though pithy and concise, is amazingly comprehensive and takes the reader through all major aspects of the human gut microbiome and microbiome science, such as the promise and limitations of germ-free animal models. Nor does Dr Shanahan shy away from the difficult topics; his discussion of 'What's normal?' is especially important and will, hopefully, disabuse those encouraged to have their microbiome profiled of the clinical value of these expensive tests. He also emphasizes the limitations of the many studies describing associations between a particular microbial profile and a disease state and reminds us that while revealing exciting avenues for future research, very few of these studies translate into causation.

What follows is a wonderful overview of the gut microbiome but, more importantly, the reader will reluctantly put the book down aware of the real facts and of all that remains to be learned.

Eamonn M.M. Quigley MD, FRCP, FACP, MACG, FRCPI, MWGO
Lynda K. and David M. Underwood Center for Digestive Disorders
Houston Methodist Hospital and Weill Cornell Medical College,
Houston, Texas, USA

Glossary of terms

Axenic: free from other live organisms – usually used in same context as gnotobiotic.

Commensal: describes the relationship between two organisms, one of which benefits while the other is unaffected. Commonly used to describe indigenous organisms that are harmless or mutually beneficial to the host.

Conventional: a control animal of the same genetic background, maintained on the same diet but colonized in an open (non-germ-free) environment.

Conventionalized: a formerly germ-free animal which has been associated with the microbes of conventional controls (but may not necessarily become colonized with a microbiota identical in composition to that of the controls).

Defined microbiota: a gnotobiotic organism maintained in an isolator and deliberately associated with one or more defined microbes.

Germ-free (GF): an organism with no live microbes in or on it.

Gnotobiotic: an organism with a known microbiota which may be germ-free or otherwise; from the Greek, *gnotos* – well known and *biota* – all of life.

Metagenome: the collection of genomes from multiple individuals within an environment and/or sample.

Microbiota: the population of organisms in a particular niche.

Microbiome: the collective genome (complete set of genes) of a given microbiota. Commonly, the terms microbiota and microbiome are used interchangeably.

Pathobiont: an organism that is normally non-threatening and part of the host's natural flora, but has the potential to cause disease under certain conditions.

Pathogen: any organism causing disease [Note: there is no such thing as a 'good' or 'bad' microbe – any microbe in the wrong place or at the wrong time can behave like a pathogen].

Specific pathogen free (SPF): laboratory mice are usually SPF meaning that a specific list of murine pathogens have been screened and excluded.

Taxon (pl. taxa): one or more organisms that are accepted as forming a unit and will usually be given a specific scientific name and ranking.

List of abbreviations

AADC: aromatic amino acid decarboxylase

ACTH: adrenocorticotrophic hormone

AHR: aryl hydrocarbon receptor

AMR: antimicrobial resistance

CRISPR: clustered regularly interspaced short palindromic repeats

DNA: deoxyribonucleic acid

ECC: enterochromaffin cells

FDA: Food and Drug Administration

FMT: fecal microbial transplant

GF: germ-free

GMO: genetically modified organism

HGT: horizontal gene transfer

HMO: human milk oligosaccharide

IBD: inflammatory bowel disease

IBS: irritable bowel syndrome

IgA: immunoglobulin A

LBP: live biotherapeutic product

LPS: lipopolysaccharide

NCCD: non-communicable chronic disease

NEC: necrotizing enterocolitis

PPI: proton pump inhibitors

RNA: ribonucleic acid

SCFA: short-chain fatty acids

SIBO: small intestinal bacterial overgrowth

SPF: specific pathogen free

TLR: Toll-like receptor

WAT: white adipose tissue

WHO: World Health Organization

1 Introduction

Gastroenterology

HEALTHCARE

A convergence of microbiome science with precision medicine promises new microbiome-based diagnostics, treatments and preventive measures for many common disorders. Success stories in microbiome science include the identification of *Helicobacter pylori* as a cause of peptic ulcer disease and stomach cancer; the treatment of recurrent *Clostridioides difficile* infection with fecal microbiota transplantation; the clarification of how gut microbes contribute to human nutrition and metabolism; and the demonstration that the microbiome can be mobilized to enhance therapeutic responses to cancer immunotherapy. The pace of research has been greatly facilitated by molecular techniques for identifying the diversity of microbes in and on the human body without the need to culture each organism. The microbiota has been linked with the risk of developing many chronic inflammatory and metabolic disorders and has been used to predict responsiveness to dietary constituents and other interventions. Although major gaps in knowledge persist, this book is about the known facts.

Scale

The scale of microbial life linked with humans is vast.
- There are as many microbes in and on the adult body as there are human cells.
- Microbial genes greatly outnumber human genes by orders of magnitude.[1,2]
- Humans are host to over 100 trillion microbes, most of which are in the gut.
- The three domains of life are represented in the gut: bacteria, archaea, and eucaria (including yeasts and protozoa) along with viruses.
- The human gut microbiota is estimated to include 500–1000 bacterial species; each bacterial strain has about 2000 genes.
- In contrast to the human genome, the microbiome – often referred to as the 'other genome' – is tractable; hence the prospect of microbiota-directed therapies.
- The gut microbiome represents a window on individual lifestyle and environmental exposure and is a biomarker of risk of disease.

Characteristic features of the human microbiome are summarized in Table 1.1.

TABLE 1.1

Characteristics of the human gut microbiome[3-18]

Feature	Comment
Evolution	The human microbiome is not random, it has been selected by evolution over eons to contribute to human physiology including digestion, metabolism and immunity. In addition, strains for most bacterial species diversified in parallel with their human host (co-diversification) as humans spread across the globe
Assembly	Colonization occurs rapidly at birth by vertical mother-to-neonate transmission which is influenced by mode of delivery (vaginal or section), followed by horizontal transmission from the environment and contact with other humans, surfaces and animals
Individuality	The microbiome is unique to each individual host with inter-individual variation occurring at species and strain levels, but much less variability at phylum level (two phyla, Bacteroidetes and Firmicutes, account for about 90% of the species in the large bowel)
Plasticity	The plasticity of the microbiome is maximal during the first 3 years of life while it is assembled and particularly in the earliest weeks and months of infancy but diminishes progressively to become relatively stable in adulthood in healthy individuals
Adaptability	On a background of relative stability, the microbiome varies in metabolic behavior and composition with diet and other lifestyle variables
Resilience	Disruption of the microbiota with antibiotics is usually followed by a return to normal, but some strains may be eliminated, especially after repeated or prolonged antibiotic exposure, with greater impact in infancy

CONTINUED

TABLE 1.1 CONTINUED

Characteristics of the human gut microbiome[3-18]

Feature	Comment
Circadian rhythm	Microbial clock genes code for oscillation in preparation for cycles of environmental change. Although gut microbes are not exposed to a light cycle, they are responsive to fluctuating internal cues such as nutrient availability and hormonal and immune signals from the host
Body-site specificity	Humans have more than one microbiome; the microbiota differs at different body sites and in many instances at different niches within the same organ
Variance	Although most (>80%) of the variance in the human microbiome is still unaccounted for, almost every environmental exposure and lifestyle variable can modify the composition and function of the microbiota whereas genetics appears to be a relatively minor contributor to microbiota variance

Spatial arrangement

Although the fecal microbiota is commonly sampled as a convenient surrogate for colorectal microbiota, the composition and diversity of the gut microbiota varies longitudinally from mouth to anus and cross-sectionally from lumen to mucosal surface (Figure 1.1).[3]

Local factors influencing the composition and density of the microbiota in each niche include availability of nutrients, pH, bile tolerance, host-derived antimicrobial factors, and oxygen tension.[4] The most abrupt transition is across the ileo-caecal valve where microbial numbers increase several log-fold.[5] Direct contact with the mucosal epithelium by microbes is restricted, and this is achieved in different ways at different sites. Unlike the colonic mucus which is bi-layered with only the outer layer containing bacteria, the small intestine, which is adapted for nutrient absorption, is covered by a single, incomplete layer of mucus protected not by a physical barrier but by antimicrobial factors such as regenerating islet-derived protein gamma.

Introduction

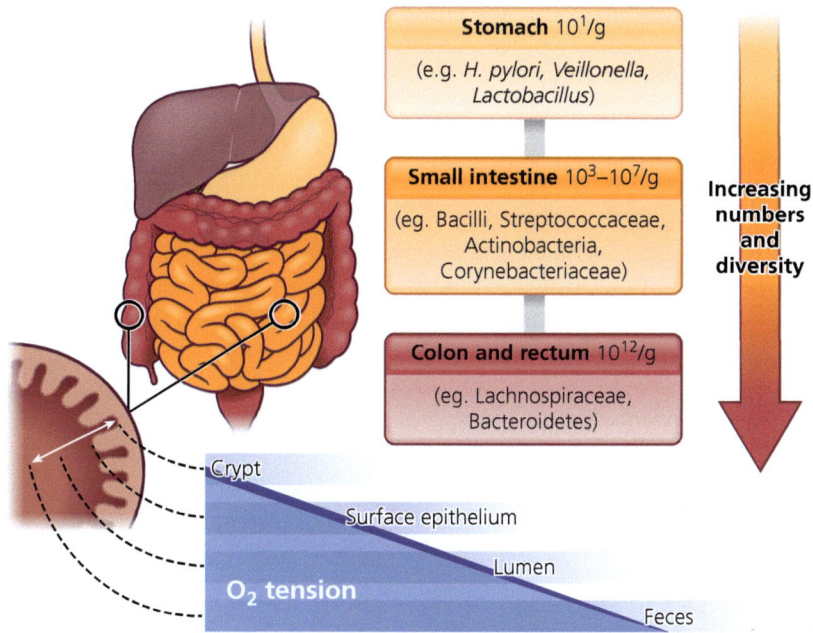

Figure 1.1 Variation in microbiota composition and density along the long axis of the gut and cross-sectionally from lumen to mucosal surface.

Protection of a microbial niche within the gut from overgrowth or infection is also mediated by self-regulatory microbe-microbe signaling. Interactions among the microbial community include predator-prey relationships, cross-feeding, scavenging and mutualistic behavior. Bacteria produce various types of quorum-sensing molecules to re-route metabolic pathways, enter a stationary phase, or, in the case of spore-forming bacterial organisms, initiate sporulation. Many bacteria also elaborate effector peptides (bacteriocins) with antimicrobial activity which may act as quorum sensors and may also kill unrelated strains. Some of the signaling molecules used by bacteria also engage with eucaryotic cells of the host, including the immune system.

Operational arrangement

Continual bi-directional host-microbe signaling is required for homeostasis. The immune system continually samples the contents of the lumen. This is achieved by allowing a low level of translocation of microbes and their metabolites from the gut by transepithelial uptake via dendritic cells and M cells overlying lymphoid follicles. Access to the systemic circulation is limited by the mesenteric lymph node, while microbes that gain entry to the portal circulation are prevented from systemic access by the liver's gatekeeper role.

A wide diversity of microbes in the gut is important not only for resilience but also for optimal performance including maximal signaling capacity with the immune system, metabolism, and the gut-brain axis. In contrast, when the microbiota has a low diversity of microbes there is less protective competition against pathogens and limited signaling capacity with the host (Figure 1.2).

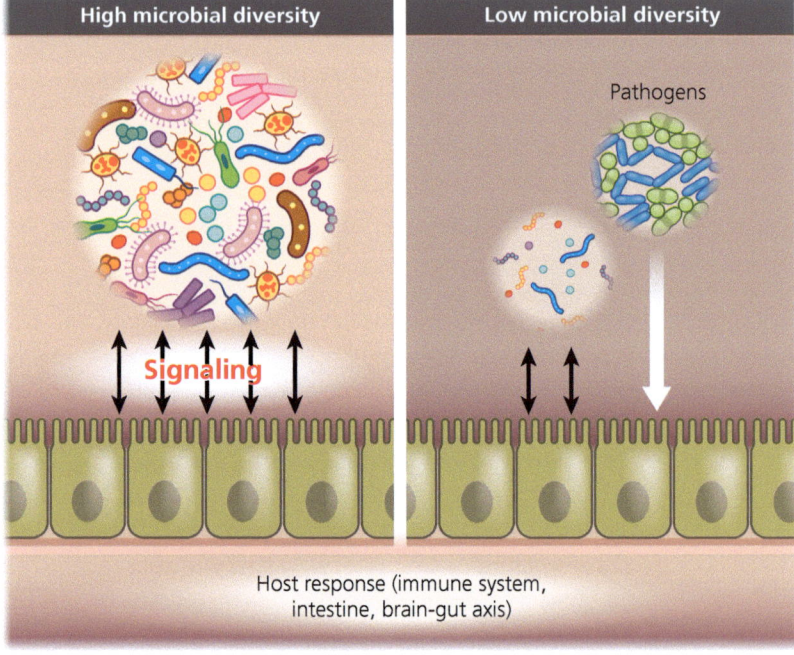

Figure 1.2 A healthy gut ecosystem requires a highly diverse microbiota to ensure resilience and performance. This includes host-microbe bi-directional signaling and specialized sites for sampling of the lumen by the host immune system.

The complexity of host-microbe signaling is exemplified by the reciprocal interaction of gut microbes with bile acids and by the expanded repertoire of signaling cascades created when microbes metabolize dietary fiber to short-chain fatty acids (SCFA),[6-7] as shown in Figure 1.3.

Temporal changes

The gut microbiota changes over time as the host ages.[8-10] It is assembled during the first three years – a critical period during which the immune system matures and is educated to distinguish danger from non-danger by exposure to harmless microbes in the immediate environment. It is also a time for maturation of metabolic pathways and neuroendocrine development under the influence of the microbiota.

Disturbances of the microbiota during this early period (for example, by exposure to antibiotics) is a risk factor for later development of immune-allergic and metabolic disorders such as obesity[11-13] (Figure 1.4).

Figure 1.3 A network of connectivity: the microbiota engages with gastrointestinal physiology at several levels. This includes modulatory effects on the enterohepatic circulation of bile acids which undergo deconjugation and conversion from primary to secondary bile acids by the microbiota and which, in turn, have direct and indirect effects (by upregulating REGIIIγ) on the composition of the microbiota. In addition, the end products of microbial digestion of dietary fiber generate short-chain fatty acids – acetate, propionate and butyrate – which nourish the gut epithelium and have pleiotropic effects on the gut-brain axis, liver metabolism, and the immune system.

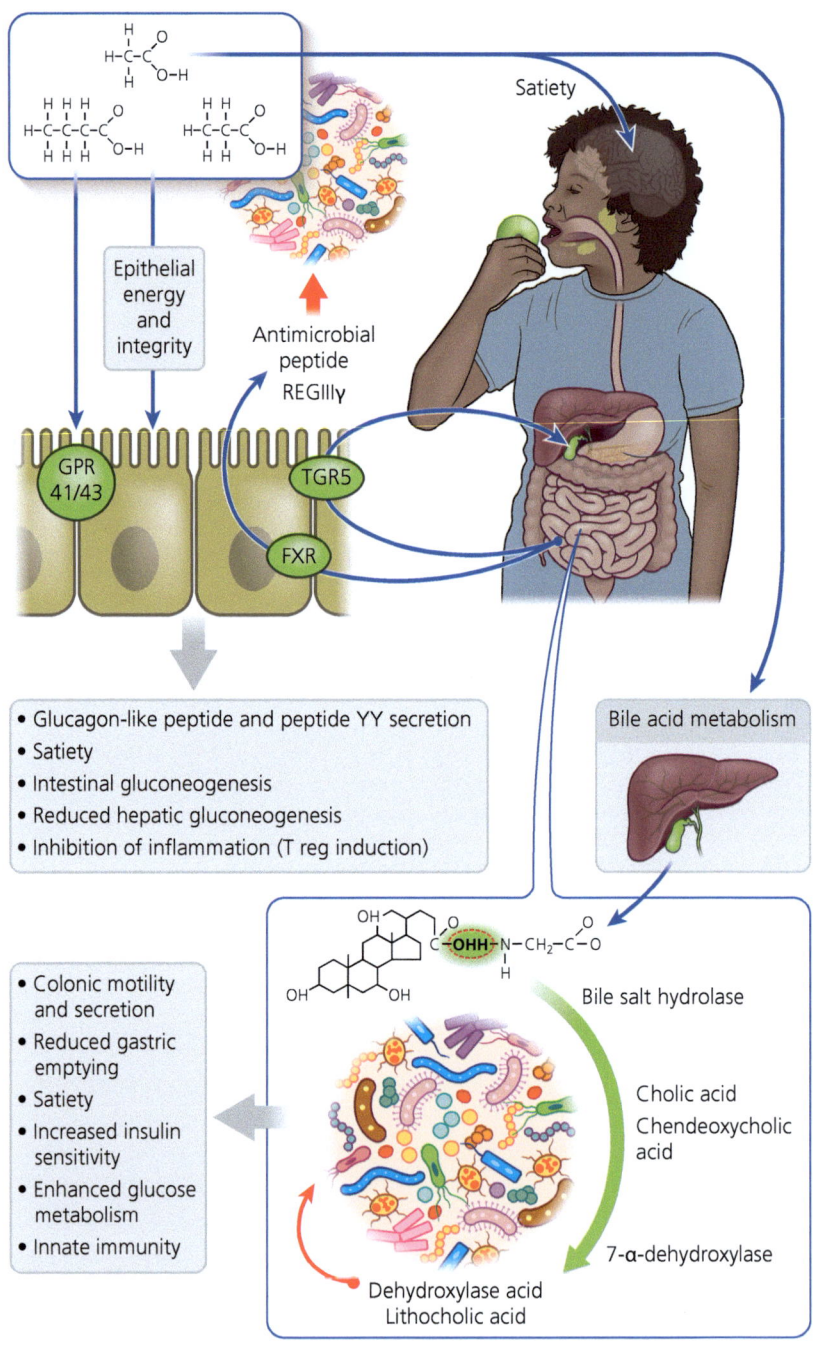

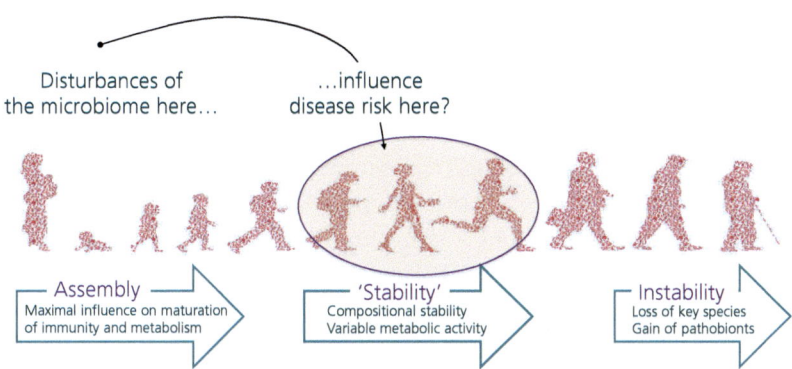

Figure 1.4 One of the main influences on the composition of the gut microbiome is the age of the host. Disruption of the microbiota during the earliest phase of life is a risk factor for development of immune, metabolic and possibly other chronic disorders, whereas disruptions of the microbiota at the other extreme of life poses an increased risk of infections, such as *Clostridioides difficile*-associated disease.

Key points – introduction

- A convergence of microbiome science with precision medicine promises new microbiome-based diagnostics, treatments, and preventive measures for many common disorders.
- An estimated 100 trillion microbes inhabit the adult body, most of which reside in the gut.
- Almost every environmental exposure and lifestyle variable can modify the composition and function of the microbiota.
- The composition and diversity of the gut microbiota varies longitudinally from mouth to anus and cross-sectionally from lumen to mucosal surface.
- A wide diversity of microbes in the gut is important not only for resilience but also for optimal performance including maximal signaling capacity with the immune system, metabolism, and gut–brain axis.
- Disruptions of the microbiota during the earliest phase of life is a risk factor for development of immune, metabolic, and possibly other chronic disorders.

References

1. Walker AW, Hoyles L. Human microbiome myths and misconceptions. *Nat Microbiol.* 2023;8:1392–1396.
2. Shanahan F, Hill C. Language, numeracy and logic in microbiome science. *Nat Rev Gastroenterol Hepatol.* 2019;16:387–388.
3. Miller BM, Liou MJ, Lee JY, Bäumler AJ. The longitudinal and cross-sectional heterogeneity of the intestinal microbiota. *Curr Opin Microbiol.* 2021;63:221–230.
4. Lee JY, Tsolis RM, Bäumler AJ. The microbiome and gut homeostasis. *Science.* 2022;377(6601):eabp9960.
5. Shanahan F, O'Toole PW. Host-microbe interactions and spatial variation of cancer in the gut. *Nat Rev Cancer.* 2014;14(8):511–512.
6. Shanahan F, van Sinderen D, O'Toole PW, Stanton C. Feeding the microbiota: transducer of nutrient signals for the host. *Gut.* 2017;66(9):1709–1717.

7. Larabi AB, Masson HLP, Bäumler AJ. Bile acids as modulators of gut microbiota composition and function. *Gut Microbes*. 2023;15(1):2172671.
8. Yatsunenko T, Rey FE, Manary MJ, et al. Human gut microbiome viewed across age and geography. *Nature*. 2012;486:222–227.
9. Roswall J, Olsson LM, Kovatcheva-Datchary P, et al. Developmental trajectory of the healthy human gut microbiota during the first 5 years of life. *Cell Host Microbe*. 2021;29:765–776.
10. Ghosh TS, Shanahan F, O'Toole PW. The gut microbiome as a modulator of healthy ageing. *Nature Reviews Gastroenterol Hepatol*. 2022; 19: 565–584.
11. Cox LM, Blaser MJ. Antibiotics in early life and obesity. *Nat Rev Endocrinol*. 2015; 11: 182–190.
12. Litvak Y, Bäumler AJ. Microbiota-nourishing immunity: a guide to understanding our microbial self. *Immunity*. 2019;51:214–224.
13. Ansaldo E, Farley TK, Belkaid Y. Control of immunity by the microbiota. *Annu Rev Immunol*. 2021; 39:449–479.
14. Carmody RN, Sarkar A, Reese AT. Gut microbiota through an evolutionary lens. *Science*. 2021;372:462–463.
15. Suzuki TA, Fitzstevens JL, Schmidt VT, et al. Codiversification of gut microbiota with humans. *Science*. 2022;377:1328–1332.
16. Browne HP, Neville BA, Forster SC, Lawley TD. Transmission of the gut microbiota: spreading of health. *Nat Rev Microbiol*. 2017;15:531–543.
17. Zhao E, Tait C, Minacapelli CD, Catalano C, Rustgi VK. Circadian rhythms, the gut microbiome, and metabolic disorders. *Gastro Hep Advances*. 2022;1:93–105.
18. Shanahan F, Ghosh TS, O'Toole PW. Human microbiome variance is underestimated. *Curr Opin Microbiol*. 2023;73:102288.

2 Life without microbes

Gastroenterology

HEALTHCARE

The potential value of germ-free (GF) animals was considered over a century ago by Louis Pasteur but it was not until the 1940s that GF technology was perfected.[1-3]

GF technology is one of the main strategies for studying host-microbe interactions, particularly to establish causal relationships. Causality may be inferred when colonization of a GF animal with one or more microbes from a donor recapitulates the donor phenotype in the recipient.[4]

Experimental studies

Methodologies. For logistical reasons, the majority of experimental studies today are performed with GF rodents but a wide range of other small (e.g., fruit fly, zebrafish) and larger (piglets, chickens) animals have been successfully raised germ free.[1,5-8]

Germ-free colonies have been developed using various techniques. Usually, a GF colony of rodents is initiated by delivering neonates via cesarean section which are then reared in an aseptic isolator. Subsequent generations of GF animals are interbred, and mothers give birth naturally in the isolator. Commercially available GF animals generated in this way are transported in a sterile container to research laboratories. Less commonly, embryos may also be transferred into a pseudo-pregnant GF mother. Once established, the colony is maintained in aseptic isolators in a GF unit where food, water, and bedding are sterile, and are regularly tested to ensure the integrity of the GF state.

While GF technology is reliable and considered to be the gold standard, researchers have also used microbial depletion with antibiotics as an alternative strategy because it is inexpensive and does not require specialized equipment and facilities. Moreover, it is applicable to animals of any genotype, whereas with GF animals, new genotypes must be rederived. The obvious disadvantages of microbial depletion include the confounding effects of antibiotics on host cells and the potential for selective overgrowth of undesirable organisms.[9]

Germ-free animals vs conventionally colonized animals.
GF animals tend to have a longer lifespan and reduced rate of spontaneous cancers than control animals but remain more

susceptible to infections which may arise from a breach of their GF conditions.

Germ-free animals also exhibit considerable nutritional and metabolic differences from control or conventionally colonized animals.[10–13] GF animals tend to be leaner and must consume 20–30% more dietary calories to maintain the same body weight. Thus, the microbiota is a net contributor to host nutrition. In addition, the microbiota normally metabolizes dietary fiber to short-chain fatty acids (SCFA) which have pleiotropic effects on host metabolism, immunity and brain function (see Figure 1.3, page 18).

The gastrointestinal features of GF mice and those of conventionally colonized animals are compared in Figure 2.1.

However, it is noteworthy that animals raised germ-free have structural and functional abnormalities in all organs that have a microbiota, including the skin and all mucosal surfaces. Moreover, because the microbiota signals to extra-intestinal organs, numerous biochemical anomalies have been reported in almost all organ systems in GF animals, most notably the brain.[14,15]

Changes in the brain include:
- altered stress responses (increased adrenocorticotrophic hormone [ACTH] and corticotrophin release with altered glucocorticoid receptor expression)
- increased blood–brain barrier permeability
- altered microglial function
- altered expression of numerous neurotransmitter agonists and receptors.

GF mice also exhibit reduced anxiety-like behavior and altered sociability and cognition. These changes are, of course, not solely attributable to absence of gut microbiota-brain signaling but may also relate to absence of microbial signals from other body sites.

Caveats to interpreting germ-free animal models
Germ-free animal models are not the same as an animal model with only microbes removed. The absence of microbes leads to structural, functional and behavioral developmental and other anomalies that cannot be ignored.

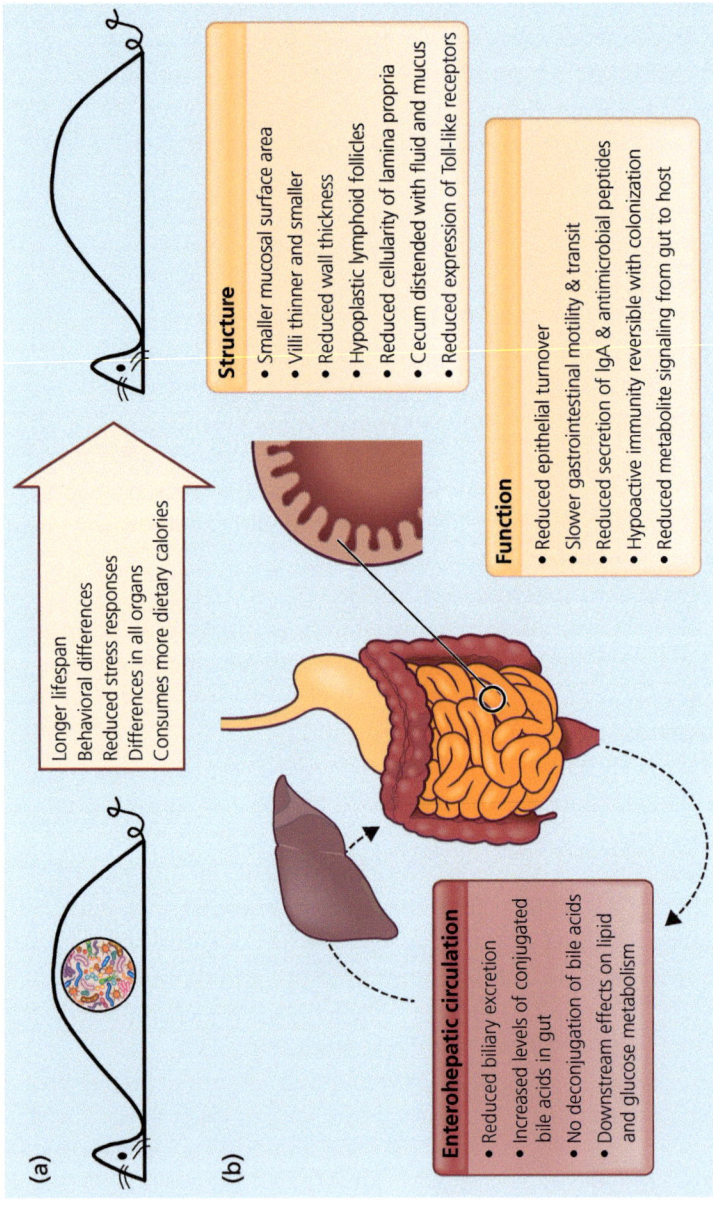

Figure 2.1 Germ-free vs conventionally colonized animals: (a) general or systemic differences; (b) gastrointestinal structural, functional and hepatobiliary differences. The list is intended to be representative, not comprehensive.

Age of colonization. For comparative studies of GF and conventionalized and conventionally-raised animals, the age at which the animals are colonized is critical. For example, early colonization with a gut microbiota is required to imprint the host's immune system in a way that is not likewise achieved at later stages in life.[16] The same applies to neuroendocrine and metabolic maturation. Similarly, since GF animals have no microbial exposure in utero or postnatally, they are of limited utility for investigating the impact of microbial alterations occurring in later in life.

Role in the pathogenesis of disease. It is noteworthy that the role of the microbiota in the pathogenesis of disease may be indirect or supportive rather than causative. For example, the absence of an inflammatory disease in a GF animal model of inflammatory bowel disease is likely due to the incapacity of a hypoplastic immune system to mount an inflammatory response that is sufficient to cause disease rather than implicating a direct effect of the microbiota.

When GF animals are colonized with microbiota from humans, a complete transfer cannot be assumed; bacteria associated with modulating host physiology are strict anaerobes and particularly susceptible to loss of viability during transfer. Failure to transfer such taxa creates a donor-recipient mismatch which may enable proliferation of opportunistic organisms in recipients and confound interpretations of the regulatory role of the microbiota on host physiology and risk of disease.

Transfer of human donor microbiotas to GF animals requires sufficient numbers of donors for statistical significance; preferably individual–not pooled–samples should be used. The unit of statistical inference relates to the number (n value) of donors of samples and not to the number of replicates (recipients). Failure to appreciate this may be behind the many reports exaggerating–rather than establishing–a role for the microbiota in several human disorders.[4]

Key points – life without microbes

- Germ-free (GF) technology is one of the main strategies for studying host–microbe interactions, particularly to establish causal relationships.
- Causality is inferred when colonization of a GF animal with one or more microbes from a donor recapitulates the donor phenotype in the recipient.
- Comparisons of GF and non-GF controls confirm that almost every aspect of host intestinal and extra-intestinal physiology is influenced by the microbiota.
- Several caveats apply to the interpretation of studies with GF animals. Developmental anomalies of structure, function and behavior must be considered.
- Transfer of human donor microbiotas to GF animals requires multiple donors – not pooled samples – for statistically valid interpretation.
- Drug-induced depletion of bacteria an alternative strategy but is confounded by effects of antimicrobials on host cells and selective overgrowth of certain organisms.

References

1. Gordon HA. The germ-free animal. *American J Dig Dis.* 1960;5:841–867.
2. Coates ME. Gnotobiotic animals in research: their uses and limitations. *Laboratory Animals.* 1975;9:275–282.
3. Kirk RGW. "Life in a germ-free world": isolating life from the laboratory animal to the Bubble Boy. *Bull Hist Med.* 2012; 86: 237–275.
4. Walter J, Armet AM, Finlay BB, Shanahan F. Establishing or exaggerating causality for the gut microbiome: lessons from human microbiota-associated rodents. *Cell.* 2020;180:221–232.
5. Uzbay T. Germ-free animal experiments in the gut microbiota studies. *Curr Opinion Pharmacol.* 2019;49:6–10.
6. Grover M, Kashyap PC. Germ free mice as a model to study effect of gut microbiota on host physiology. *Neurogastroenterol Motil.* 2014; 26(6): 745–748.

7. Martin R, Bermúdez-Humarán LG, Langella P. Gnotobiotic rodents: an in vivo model for the study of microbe-microbe interactions. *Front Microbiol.* 2016;7:409.
8. Kamareddine L, Najjar H, Sohail MU, Abdulkader H, Al-Asmakh M. The microbiota and gut-related disorders: insights from animal models. *Cells.* 2020. 9(11):2401.
9. Kennedy EA, King KY, Baldridge MT. Mouse microbiota models: comparing germ-free mice and antibiotics treatment as tools for modifying gut bacteria. *Front Physiol.* 2018;9:1534.
10. Thompson GR, Trexler PC. Gastrointestinal structure and function in germ-free or gnotobiotic animals. *Gut* 1971;12:230–235.
11. Sayin SI, Wahlström A, Felin J, Jäntti S, et al. Gut microbiota regulates bile acid metabolism by reducing levels of tauro-beta-muricholic acid, a naturally occurring FXR antagonist. *Cell Metab.* 2013;17:225–235.
12. Ridaura VK, Faith JJ, Rey FE, et al. Gut microbiota from twins discordant for obesity modulate metabolism in mice. *Science.* 2013; 341(6150):1241214.
13. Goodman AL, Kallstrom G, Faith JJ, et al. Extensive personal human gut microbiota culture collections characterized and manipulated in gnotobiotic mice. *Proc Natl Acad Sci USA.* 2011; 108:6252–6257.
14. Diaz Heijtz R, Wang S, Anuar F, et al. Normal gut microbiota modulates brain development and behavior. *Proc Natl Acad Sci USA.* 2011; 108(7):3047–3052.
15. Luczynski P, McVey Neufeld KA, Oriach CS, Clarke G, Dinan TG, Cryan JF. Growing up in a bubble: using germ-free animals to assess the influence of the gut microbiota on brain and behavior. *Int J Neuropsychopharmacol.* 2016;19(8):pyw020.
16. El Aidy S, Hooiveld G, Tremaroli V, Bäckhed F, Kleerebezem M.The gut microbiota and mucosal homeostasis: colonized at birth or at adulthood, does it matter? *Gut Microbes.* 2013;4:118–124.

Gastroenterology

3 What's normal?

HEALTHCARE

A universally accepted normal or healthy microbiome has not been defined, but certain generalizations can be made.[1]
- The more stringent the definition of normality, the rarer it becomes.
- Several configurations of the microbiota are consistent with health, and no single microbiome configuration is essential for health.
- Normal is not the same as healthy (it was once normal for almost all of the population to have *Helicobacter pylori* in the stomach, but one would not refer to that as healthy, given the causative role of *H. pylori* in peptic ulceration and stomach cancer).
- Health cannot merely be the absence of disease if many seemingly healthy individuals harbor microbial risk factors for disease, such as cancer of the stomach or colon.
- Normality is not synonymous with the average or typical if most people in the population are overweight or obese (Box 3.1).

Nor can 'normal' be the ideal state or even desirable because these concepts vary with context. A microbiome adapted to maximal extraction of calories from food may be an asset in a famine-stricken country but a liability in an obesogenic environment. Equally, a microbiome that is beneficial for a full-term neonate may represent a risk factor for infection in a preterm baby born before the immune system, the mucosal barrier and the blood-brain barrier have fully developed.

Knowns and unknowns

Population-based studies have shown that most of the variance in human microbiome composition has not been accounted for (Figure 3.1).[2-4] Similarly, in cohort studies of patients with inflammatory bowel disease (IBD), most of the variance in

---

BOX 3.1

What's normal?

Normal ≠ Healthy

Healthy ≠ Absence of disease

Normal ≠ 'Average' or 'Typical'

Normal ≠ 'Ideal' or 'Desirable'

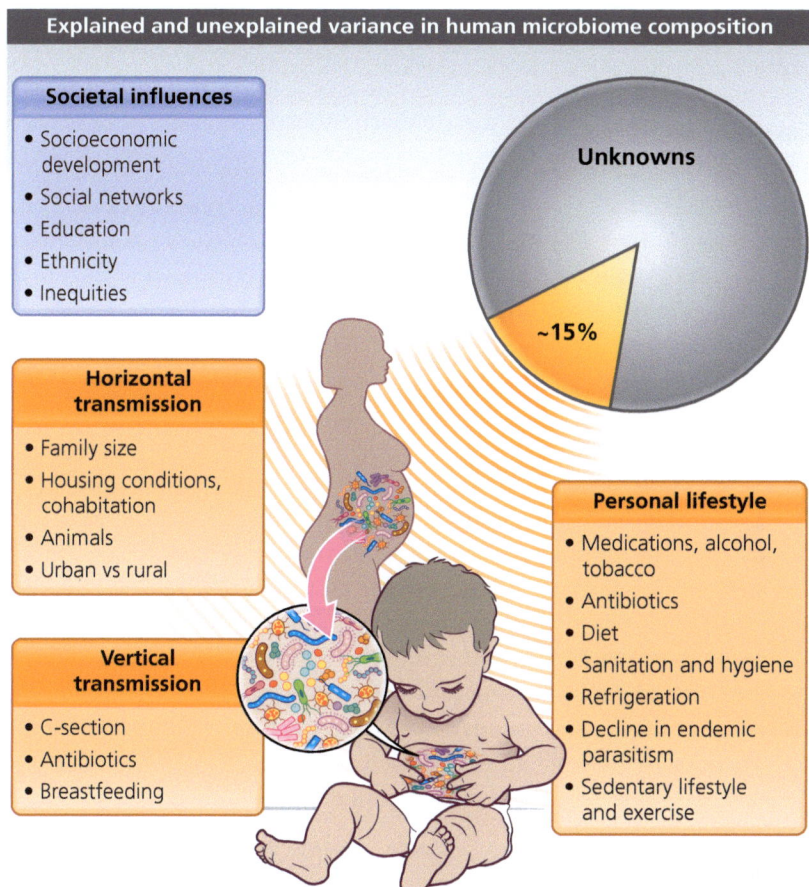

Figure 3.1 Overview of the known environmental and lifestyle modifiers of the human gut microbiota composition and function.

microbiome composition remains unexplained.[5] While host genetics have an influence, the microbiome is modified mainly by environmental and lifestyle factors. Many of these modifying influences are obvious and well established whereas others operate at a more complex sociocultural or socioeconomic level.[6,7]

For example, ethnicity, geography and many of the social determinants of health including social networks, marginalization, aloneness, discrimination, poverty and education have overlapping, interactive effects on the microbiome.

Sex and gender

One of the most important determinants of human difference is sex and gender; both have an impact on microbiome composition and function but have been underreported. Sex is biologically determined while gender reflects sociocultural influences that interact with biology and behavior.[8,9]

Gut microbes affect estrogen metabolism and blood estrogen levels by deconjugating estrogen after it has been conjugated in the liver and excreted into the bile (estrobolome). Conjugated estrogen is excreted in the feces but is available for reabsorption after microbial deconjugation. By altering estrogen levels, the microbiome may influence the risk of hormone-driven cancers in the endometrium, ovary, prostate, and breast and may also affect menstruation and fertility. Androgens, such as testosterone, also undergo metabolism by gut microbes and recirculate from gut to blood to liver and back. The microbiota in premenopausal women seems to be more diverse than that of men and some investigators have suggested that compositional differences may contribute to the sex bias in autoimmunity, which is more common in women, and to cardiovascular disease, which has a male bias, and might have a role to play in the longer lifespan of women.

Relationship to disease – cause, consequence or contributory?

Alterations in microbiome composition or function have been linked with a wide range of chronic disorders, but few aside from those linked with single microbes such as *Clostridioides difficile* or *H. pylori* have a proven cause-and-effect mechanism. More complex microbiome-disease relationships may contribute to chronic inflammatory and metabolic disorders, where the risk of developing disease is best considered as a function of the opposing influences of disease-associated taxa against protective taxa, with the emergence of overt disease being context dependent (Figure 3.2).

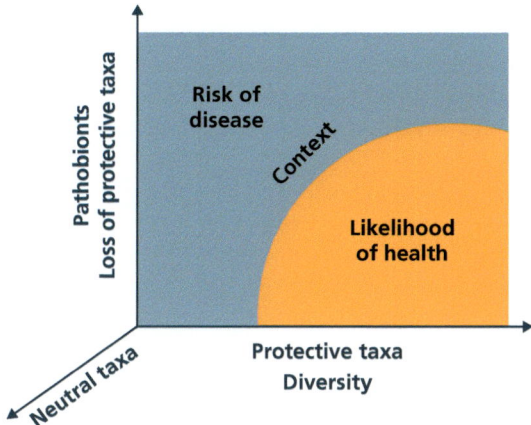

Figure 3.2 The relationship between the microbiome and risk of disease as a function of protective and disease-associated (pathobiont) taxa, with the emergence of overt disease being context dependent.

For many disorders, some of the reported microbiome alterations are consequential not causal.[10] For example, changes in mucosal oxygen tension occur with all forms of mucosal inflammation and can be expected to change the microbiota with a bloom of proteobacteria such as *Escherichia coli*.[11] Drug treatment of disease may also alter the microbiota; a noteworthy example is the influence of biguanide therapy (metformin) in type 2 diabetes mellitus.[12] In the case of autism, the microbiota alterations appear to be secondary to habitual dietary preferences.[13]

Regardless of whether the microbiota alterations are causal or consequential, they may modify the disease activity and thus merit therapeutic attention (Table 3.1). A common example is use of laxatives and/or topical antibiotics such as neomycin or rifaximin in hepatic encephalopathy to reduce the uptake into the portal circulation of ammonia and other toxic microbial metabolites which would normally be metabolized by a non-diseased liver. Similarly, the primary problem in inherited familial dysautonomia is progressive neuronal degeneration, but when the gastrointestinal tract is involved there are microbiome disturbances which exacerbate the neurodegeneration.[14] In more common disorders, such as chronic

TABLE 3.1

Host-microbe interactions in human disease*

Relationship	Example
Causal	• *Clostridioides difficile*-associated disease
	• *Helicobacter pylori*-associated disease
Consequential (at least in part)	• Change in O_2 tension (mucosal inflammation)
	• Ectopic ingress of oral microbiota (IBD, colon cancer, aging)
	• Drug treatment with metformin (type 2 diabetes mellitus)
	• Dietary preference (autism)
Contributory	• Liver failure & hepatic encephalopathy
	• Familial dysautonomia
	• Chronic kidney disease
	• Cancer immunotherapy

*Examples are representative, not comprehensive.
IBD, inflammatory bowel disease.

kidney disease, the gut microbiota may also be contributory because of failure to excrete toxic microbial metabolites, a problem that may be responsive to dietary modification.[15,16]

Improved responses to existing treatments and dietary regimens may also be achieved by modulating the microbiome as shown by encouraging evidence linking the microbiome with responsiveness to cancer immunotherapy.

Spectrum of gut microbiomes

Definitions of a normal or a healthy microbiome are confounded by the fact that most studies on the human microbiome have been based on samples from affluent, modern, industrialized societies of the Western world.[4,17] However, the composition and functional capacity of the microbiomes of humans living in non-industrialized regions of the globe differ markedly from those of people in the Western world (Figure 3.3).

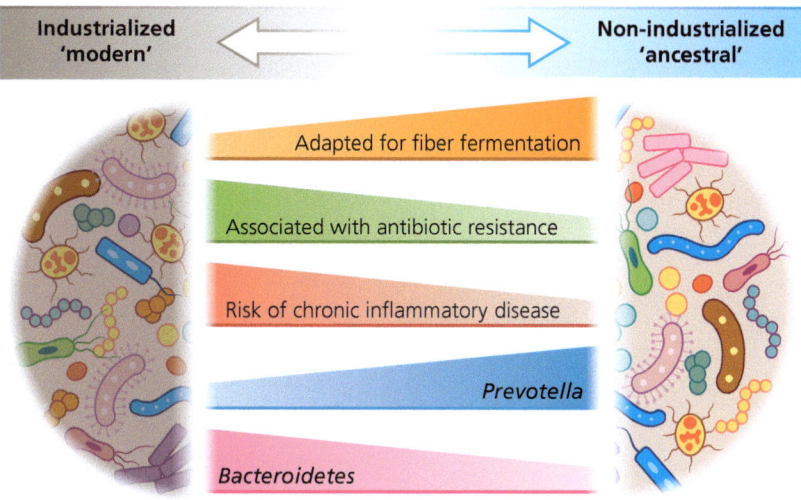

Figure 3.3 Spectrum and features of industrialized to non-industrialized microbiomes.

Studies of preindustrial microbiomes obtained from ancient feces that have been naturally preserved in frozen, desiccated, or salted conditions show that non-industrialized human gut microbiomes from developing countries resemble those of ancestral times. The modern industrialized human gut microbiome appears to have diverged from an ancestral state due to modernization, including diet and other factors. It is, however, noteworthy that non-industrialized and industrialized microbiotas are each heterogeneous while also representing distal ends of a spectrum with a gradient of transitional microbiomes.

Lessons from a minority microbiome

Mass migrations and multiethnicity, increasingly a feature of modern urban societies, represent a challenge to the concept of a normal microbiome. This is underlined by the unique microbiomes of Irish Travellers – an ethnic minority with a distinct history, culture, and language, who account for about 1% of the population of Ireland.[18] Despite living in a modern industrialized country, this ethnic minority has retained a non-industrialized type of microbiome. Although dietary differences account for the distinctive microbiomes of many ethnic minorities, they do not explain the unique microbiome of the Irish Travellers which is mainly attributable to a lifestyle akin to that of a bygone age: large families, living in close quarters and in proximity to their domestic animals, mainly horses.

 Key points – what's normal?

- There is no universally accepted definition of a normal or healthy microbiome.
- In population-based studies, most of the variance in the composition of the microbiome is unexplained.
- Host genetics have an influence but environmental, lifestyle and sociocultural factors have a major role in shaping the human microbiome.
- Mass migrations and multiethnicity, increasingly a feature of modern urban societies, represent a challenge to the concept of a normal microbiome.
- Sex and gender influences – previously underreported – have emerged as important determinants of human microbiome differences.
- Most microbes are harmless; their relationship to disease depends on context.
- Alterations in the microbiome associated with many disorders are not causal but result from the disease or its drug treatment.

References

1. Shanahan F, Ghosh TS, O'Toole PW. The healthy microbiome – what is the definition of a healthy gut microbiome? *Gastroenterology*. 2021;160:483–494.
2. Falony G, Joossens M, Vieira-Silva S, et al. Population-level analysis of gut microbiome variation. *Science*. 2016;352:560–564.

3. Zhernakova A, Kurilshikov A, Bonder MJ, et al. Population-based metagenomics analysis reveals markers for gut microbiome composition and diversity. *Science*. 2016;352:565–569.
4. Shanahan F, Ghosh TS, O'Toole PW. Human microbiome variance is underestimated. *Curr Opin Microbiol*. 2023;73:102288.
5. Clooney AG, Eckenberger J, Laserna-Mendieta E, et al. Ranking microbiome variance in inflammatory bowel disease: a large longitudinal intercontinental study. *Gut*. 2021;70:499–510.
6. Sarkar A, Harty S, Johnson KVA, et al. Microbial transmission in animal social networks and the social microbiome. *Nat Ecol Evol*. 2020;4:1020–1035.
7. Amato KR, Arrieta MC, Azad MB, et al. The human gut microbiome and health inequities. *Proc Natl Acad Sci USA*. 2021;118: e2017947118.
8. Rosser EC, de Gruijter NM, Matei DE. Mini-review: gut microbiota and the sex bias in autoimmunity – lessons learnt from animal models. *Front Med*. 2022;9:910561.
9. Kim YS, Unno T, Kim BY, Park MS. Sex differences in gut microbiota. *World J Mens Health*. 2020;38:48–60.
10. O'Toole PW, Ghosh TS, Goswami S, Manghi P, Segata N, Shanahan F. Translating the microbiome: what's the target? *Gastroenterology*. 2023;165:317–319.
11. Zeng MY, Inohara N, Núñez G. Mechanisms of inflammation-driven bacterial dysbiosis in the gut. *Mucosal Immunol*. 2017;10:18–26.
12. Forslund K, Hildebrand F, Nielsen T, et al. Disentangling type 2 diabetes and metformin treatment signatures in the human gut microbiota. *Nature*. 2015;528:262–266.
13. Yap CX, Henders AK, Alvares GA, et al. Autism-related dietary preferences mediate autism-gut microbiome associations. *Cell*. 2021;184:5916–5931.
14. Cheney AM, Costello SM, Pinkham NV, et al. Gut microbiome dysbiosis drives metabolic dysfunction in familial dysautonomia. *Nat Commun*. 2023; 14(1): 218.
15. Lobel L, Cao YG, Fenn K, Glickman JN, Garrett WS. Diet post-translationally modifies the mouse gut microbial proteome to modulate renal function. *Science*. 2020; 369(6510): 1518–1524.
16. Pluznick JL. The gut microbiota in kidney disease. *Science*. 2020; 369(6510): 1426–1427.

17. Sonnenburg ED, Sonnenburg JL. The ancestral and industrialized gut microbiota and implications for human health. *Nat Rev Microbiol.* 2019;17:383–390.

18. Keohane DM, Ghosh TS, Jeffery IB, et al. Microbiome and health implications for ethnic minorities after enforced lifestyle changes. *Nat Med.* 2020;26:1086–1095.

4 The changing microbiome as the host ages

Gastroenterology

HEALTHCARE

The human gut microbiome is not static but it changes over the lifetime of the host (see Figure 1.4, page 19).

A prenatal microbiome?

One of the most controversial issues in human biology is the mechanism by which fetal immunity is prepared for a microbial world. How can this be achieved without exposing the immune system to microbes before birth?

The long-held belief that the womb is sterile appeared to have been refuted when microbes were reported within the normal placenta and in fetal tissues. However, microbes apparently detected in healthy fetal tissue have been dismissed by many scientists as contaminants.[1] Although it is practically impossible to prove a negative–that something does not exist–a plausible alternative has been offered. Since the immune system can be trained by vaccinating with inactive or harmless microbial components, it seems implausible that nature would risk using live microbes for the developing immune system. Instead of a fetal microbiome, it has been proposed that the developing fetal immune system is exposed to microbial fragments and metabolites by transplacental passage from mother to baby.

The neonatal microbiome

The first determinant of the composition of the neonatal microbiome is mother-to-infant vertical transmission. Although microbes from different maternal body sites are transmitted to the infant, the maternal gut microbiome seems to be the dominant source of vertically-transmitted strains, consistent with fecal-microbial seeding during vaginal birth. Vertical transmission between the generations has implications for the evolution of the human microbiome, notably the adaptation and coevolution of host and microbe.[2-4]

The mode of birth delivery has a profound influence on the initial microbial colonization during the assembly of the neonatal microbiome (Table 4.1). The microbial consequences of the two modes of delivery diminish over time, while putative long-term health consequences are unproven.

TABLE 4.1

Influence of mode of delivery on assembly of neonatal gut microbiome

Vaginal	Cesarean section
Colonization with vaginal microbes	Colonization from maternal skin
Enriched with *Bifidobacterium*, *Escherichia*, and *Bacteroides* species	Enriched with species associated with hospital environment
More stable in first year	Less stable (less well adapted to gut)
Less heterogeneous	More heterogeneous
Mother-infant strain sharing likely	Less likely to exhibit strain sharing

Early life

The next wave of microbes to colonize the neonate arrives from various sources including paternal, hospital staff, older siblings, or from contact with household objects and domestic animals. The predominant taxa in the gut quickly change during the first month from facultative aerobes (e.g. Enterobacteriaceae) to strict anaerobes – mostly *Bifidobacterium*, *Bacteroides*, and *Clostridium*.

Over the next few years, there is a progressive increase in microbial load and complexity; the infant microbiome resembles that of an adult at 3–4 years of age.[2] During that period, the introduction of solid food and withdrawal of breastfeeding are important milestones and modifiers of the microbiome with increases in early colonizers such as Clostridiales and *Bacteroides* and a slow diminution in *Bifidobacterium* species.

Strain-level analysis shows that the prominent early colonizing strains of bifidobacteria are adapted to metabolizing human milk oligosaccharides (HMOs),[4,5] whereas strains which persist in adulthood are adapted to fermentation of non-HMO carbohydrates (Box 4.1). However, some HMO-adapted strains are thought to persist into adulthood, ready for vertical transmission to the next generation. This is consistent with reported remodeling of the gut microbiome during pregnancy, including increased abundance of bifidobacteria in late pregnancy in response to progesterone.[6,7]

> **BOX 4.1**
> **Properties of human milk oligosaccharides (HMOs)**
>
> - Heterogeneous: >200 structurally distinct HMOs described
> - Maternal-specific: each woman produces a subset of the possible structures
> - Third most abundant solid component of breast milk
> - Indigestible to the infant and reach the intestine intact
> - Influence on microbiome: bifidogenic prebiotic effect
> - Mainly digested by *Bifidobacterium* spp. but can be metabolized by other bacteria including species within the *Bacteroides* genus
> - Protect by promoting bacteria to occupy mucosal niche and act as decoy receptors and anti-adhesive antimicrobials to prevent pathogens adhering to the epithelium
> - Modulate the epithelium (growth factor effect, reduced permeability)
> - Modulate host immunity
> - Metabolites of HMOs – small chain fatty acid metabolites (acetate, butyrate, propionate and others) – mediate many of their beneficial effects on the host

Breast milk, which contains skin and enteric bacteria in addition to HMOs, nutrients, immunoglobulin A, lactoferrin, lysozyme, growth factors and other bioactives, is probably the single most important modifier of the infant gut microbiome.[8] While the benefits of breastfeeding are substantial, formula-fed babies are not necessarily at major disadvantage. The value of breastfeeding is particularly important for preterm babies.

Born too soon

Preterm infants are more likely to be exposed to antibiotics than full-term infants. Neonatal intensive care and contact with non-maternal carers increases their likelihood of acquiring strains from the hospital environment. Hospital-acquired strains may be poorly adapted to the intestinal environment and may not persist in the infant gut.

The preterm infant microbiome is highly variable and of low diversity.[9,10] The *Bifidobacterium*-predominant microbial community characteristic of full-term infants is less common in preterm babies. The shorter the gestational age, the greater is the variability and unpredictability of the microbiome. In extremely premature infants, it is uncertain if the intestinal microenvironment can support strict anaerobes. There may be a dominance of *Staphylococcus*, *Klebsiella*, *Enterococcus* or *Escherichia*, with fluctuating shifts in dominance from one genus to another.

Microbial anomalies with an underdeveloped mucosal barrier and immune system predispose preterm infants to sepsis and necrotizing enterocolitis (NEC)[11,12] (Table 4.2).

TABLE 4.2

Features of necrotizing enterocolitis (NEC)

Frequency	The most commonly acquired serious gastrointestinal complication of prematurity
	Rare after 32 weeks gestation
Etiology	Multifactorial: immature mucosal barrier function and intestinal microenvironment, underdeveloped immunity, variable microbial assembly
Pathogenesis	Complex, involving pathobiont overgrowth, epithelial injury, mucosal inflammation and microvascular insufficiency
Commonly implicated pathobionts (insufficient alone to cause NEC)	*C. perfringens*, *Klebsiella* spp.
Protective organisms	Inversely correlated with higher relative abundance of *Bifidobacterium* spp.
Treatment	Breast milk, probiotics, supportive care

Growing old with microbes

Everyone ages but not at the same rate. The microbiome is one of the determinants of healthy aging and is emerging as a biomarker of the difference between chronological age and biological age.[13] The microbiome has a reciprocal relationship with the age of the host; it deteriorates as the host undergoes age-related physiological decline, while simultaneously it contributes to the aging process and age-related diseases (Figure 4.1).

In addition, societal and social variables that shape the microbiome may be particularly relevant for older people and include aloneness, loss of contact with the external microenvironment and institutional care.

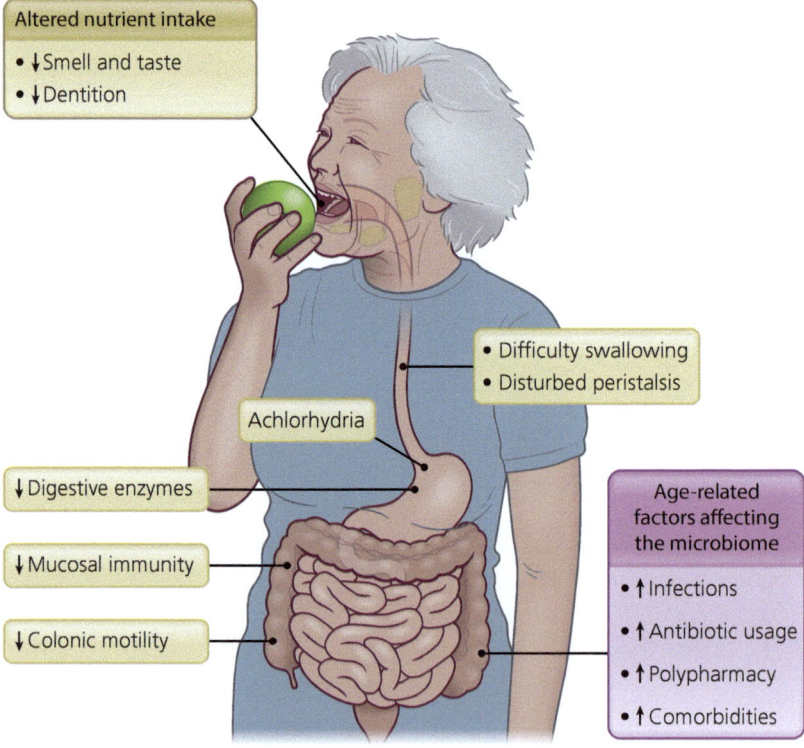

Figure 4.1 Variable physiological deterioration with aging and age-related factors that influence the microbiome.

> **BOX 4.2**
> **Social and societal modifiers of the microbiome of older people**
> - Loss of human contact
> - Living alone
> - Household pets
> - Visitors
> - Institutional or residential accommodation
> - Life indoors

Microbial changes of aging

The microbial changes associated with aging include reduced microbial load, particularly with loss of protective taxa and greater relative abundance of pathobionts (disease-associated taxa) (Box 4.2).[14] In addition, all measures of uniqueness and diversity increase with age. More importantly, the abundance of specific taxa strongly discriminates unhealthy and healthy aging trajectories in different geographic cohorts (Figure 4.2). For example, a group of species-level

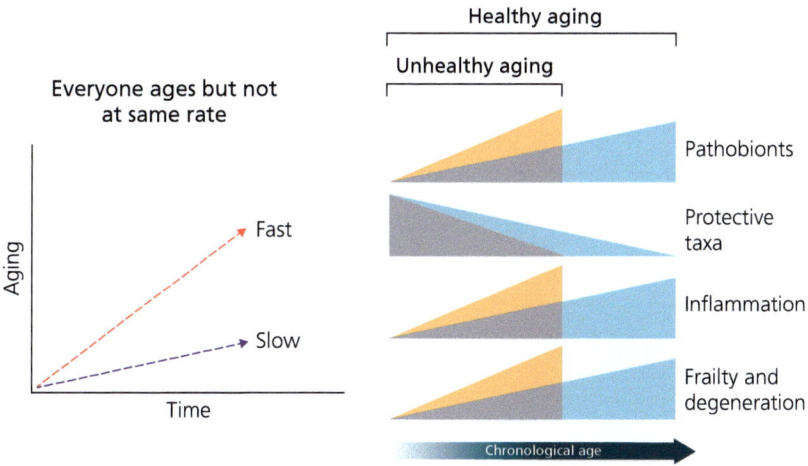

Figure 4.2 Microbe-host interactions as a contributor to healthy and unhealthy aging. Adapted from Ghosh et al. 2022.[13]

taxa devoid of any *Bacteroides* species has been reported to exhibit the strongest association with healthy aging.[14]

Delaying the age-related deterioration of the microbiome may be possible using dietary modification. For example, the introduction of a Mediterranean diet (lower consumption of red meat, dairy products and saturated fats with increased intake of vegetables, fruits, legumes, fish, olive oil and nuts) has been linked with microbiome restoration and reduced trajectory of frailty and cognitive decline.[15]

Key points – the changing microbiome as the host ages

- The concept of a live prenatal microbiome has been challenged; exposure to microbial fragments–not live microbes –occurs by transplacental passage from mother to baby.
- Mode of birth delivery (vaginal or cesarean) has a profound influence on microbial colonization of the neonate.
- Human breast milk and its multifunctional oligosaccharides represent the single most important modifier of the infant gut microbiome and are particularly important for preterm babies at risk of necrotizing enterocolitis.
- The microbiome changes with age but is also one of the determinants of healthy aging and a biomarker of biological versus chronological age.
- Deterioration of the microbiome with age is linked with increased inflammation, frailty and accelerated cognitive decline.

References

1. Kennedy KM, de Goffau MC, Perez-Muñoz ME, et al. Questioning the fetal microbiome illustrates pitfalls of low-biomass microbial studies. *Nature*. 2023;613:639-649.
2. Yatsunenko T, Rey FE, Manary MJ, et al. Human gut microbiome viewed across age and geography. *Nature*. 2012;486:222-227.
3. Roswall J, Olsson LM, Kovatcheva-Datchary P, et al. Developmental trajectory of the healthy human gut microbiota during the first 5 years of life. *Cell Host Microbe*.2021;29: 765–776.
4. Enav H, Bäckhed F, Ley RE. The developing infant gut microbiome: a strain-level view. *Cell Host Microbe*. 2022;30: 627-638.
5. Masi AC, Steward CJ. Untangling human milk oligosaccharides and infant gut microbiome. *iScience*. 2021;25(1):103542.
6. Koren O, Goodrich JK, Cullender TC et al. Host remodeling of the gut microbiome and metabolic changes during pregnancy. *Cell*. 2012; 150:470-480.
7. Nuriel-Ohayon M, Neuman H, Ziv O et al. Progesterone increases Bifidobacterium relative abundance during late pregnancy. *Cell Rep*. 2019;27:730-736.
8. Stewart CJ. Breastfeeding promotes bifidobacterial immunomodulatory metabolites. *Nat Microbiol*. 2021;6:1335-1336.
9. Healy DB, Ryan CA, Ross RP, Stanton C, Dempsey EM. Clinical implications of preterm infant gut microbiome development. *Nat Microbiol*. 2022;7:22-33.
10. Rao C, Coyte KZ, Bainter W, Geha RS, Martin CR, Rakoff-Nahoum S. Multi-kingdom ecological drivers of microbiota assembly in preterm infants. *Nature*. 2021;591(7851):633-638.
11. Duess JW, Sampah ME, Lopez CM, et al. Necrotizing enterocolitis, gut microbes, and sepsis. *Gut Microbes*. 2023;15:2221470.
12. Razak A, Patel RM, Gautham KS. Use of probiotics to prevent necrotizing enterocolitis. Evidence to clinical practice. *JAMA Pediatr*. 2021;175:773-774.
13. Ghosh TS, Shanahan F, O'Toole PW. The gut microbiome as a modulator of healthy ageing. *Nat Rev Gastroenterol Hepatol*. 2022;19(9):565-584.
14. Ghosh TS, Shanahan F, O'Toole PW. Toward an improved definition of a healthy microbiome for healthy aging. *Nat Aging*. 2022;2(11):1054-1069.
15. Ghosh TS, Rampelli S, Jeffery IB, et al. Mediterranean diet intervention alters the gut microbiome in older people reducing frailty and improving health status: the NU-AGE 1-year dietary intervention across five European countries. *Gut*. 2020;69(7):1218-1228.

5 Food and microbes

Gastroenterology

HEALTHCARE

When animals eat they feed not only themselves but also their microbes. Maintenance of a healthy microbiota requires a healthy diet. Dietary intake determines which microbes are retained, flourish or become extinct, and loss of dietary diversity leads to a reduction in gut microbial diversity. In addition, microbes are net contributors to host nutrition and provide the metabolic capacity for energy extraction from dietary fiber.[1-4]

Fiber: food for microbes

Humans can't digest dietary fiber but their microbes can (Box 5.1). Most dietary fibers can be fermented and broken down by microbes to simpler end products – short-chain fatty acids (SCFA) – in the gut.[5,6]

The most abundant SCFA in the human gut are acetate (C2), propionate (C3) and butyrate (C4). These greatly expand the nutritional and functional impact of dietary fiber on intestinal and extraintestinal physiology including immune, metabolic, neuroendocrine and brain function, and emphasize the potential scale of the adverse consequences of a fiber-deficient diet (Figure 5.1). The effects of SCFA are largely mediated though G-protein-coupled receptors which are differentially expressed in distinct subsets of epithelial, immune and endocrine cells.

Intake of dietary fiber has declined over the past century with increasing socioeconomic development and industrialization. This has been linked with an increased risk of many inflammatory and metabolic chronic disorders. However, when the content of dietary fiber is changed from low to high, the response is influenced by the baseline composition of the gut microbiota with favorable responses associated with a higher *Prevotella*/Bacteriodes ratio.[7]

Linking diet and specific microbial metabolites with disease

In addition to the generation of SCFA, other diet-microbe interactions yield important metabolites that influence the health of the host (Table 5.1).[1,2,3,8]

The microbiota and fat storage and metabolism

In addition to energy harvest from the diet, the microbiota also regulates fat storage.[2,8] The microbiota inhibits intestinal

> **BOX 5.1**
> **Properties of dietary fiber**
>
> - Heterogeneity: fiber is not a single entity – it is a collective term for a group of complex carbohydrates, most of which can be broken down to simpler end products such as short-chain fatty acids (SCFA) by gut microbes
> - Plant-derived: fibers eaten by humans are mainly from plants in which they provide structural scaffolding
> - Consist of repeating carbohydrate units (polysaccharides) of variable length and number of side chains.
> - The behavior of fibers in the gut is determined by three overlapping features – solubility, viscosity and fermentability
> - Small variations in fiber structure are associated with distinctly different effects on the gut microbiome. Different fiber structures may shift SCFA metabolites toward either butyrate or propionate
> - Fiber-specific microbial responses: microbes vary in their preference for different fiber types
> - Dose-dependent influence on the microbiome with a plateau of ~35g/day

fasting-induced adipose factor (FIAF) which is an inhibitor of lipoprotein lipase (LPL) in white adipose tissue (WAT) and thereby increases fat storage in WAT.

The microbiota may also influence body weight by promoting WAT beiging and increased energy expenditure. Beiging or browning is a progressive acquisition by WAT of properties normally associated with brown fat.

Gas production by microbes

Microbial production of gas (hydrogen H_2, carbon dioxide CO_2, and methane CH_4) in the gut is mainly from metabolism of dietary constituents, particularly carbohydrates, although intestinal mucus and glycoproteins are potential substrates. This occurs mainly in the distal small bowel and colon, whereas gas in the stomach and proximal small bowel (oxygen O_2, nitrogen N_2, and carbon dioxide

CO_2) is from air-swallowing or is produced by gastric acid interacting with pancreatic and duodenal bicarbonate (Figure 5.2).

The undernourished microbiota

Inadequate nutrition in early life causes stunting of growth, impaired cognitive and immune development, and an increased risk of infections.[9-11] These deficits persist if nutrition is not promptly restored. However, provision of the most readily available food does not solve the problem unless it also nourishes the microbiome.

A malnourished microbiota can lead to impaired development of brain, immune, and metabolic function. Since currently available nutritional interventions do not resolve the microbiome damage or the long-term consequences of malnutrition, an alternative approach has been to develop complementary foods that have a beneficial effect on the microbiome based on extensive screening in experimental animals colonized with microbes from humans. The ingredients for such foods are selected from locally available sources and encouraging

---

Figure 5.1 Range of nutritional and signaling effects of short-chain fatty acids (SCFA). Butyrate is the preferred metabolic substrate for intestinal epithelial cells. Bacteria-derived butyrate reduces epithelial O_2 consumption and stablizes hypoxia-inducible factor (HIF), a transcription factor coordinating barrier protection. SCFA also enhance mucin production by goblet cells and trigger the release of hormones from enteroendocrine cells such as glucagon-like peptide 1 (GLP-1) which leads to satiety, and peptide YY, and GLP-2. SCFA also modulate the synthesis of 5-hydroxytryptamine (5-HT or serotonin) in enterochromaffin cells (ECCs). In addition, SCFA modulate enteric neuronal afferent signaling and may also influence the brain-gut axis via the systemic circulation. Note extensions of ECCs (neuropods) are in direct contact with vagal and possibly sympathetic afferent nerve terminals. SCFA also shape mucosal immune responses by actions on dendritic cells and regulatory T cells (T reg), and enhance immunoglobulin A (IgA) production and downregulate inflammation. SCFA may gain access to the systemic circulation and directly affect metabolism and function of peripheral tissues and reach the brain and cerebrospinal fluid where they may influence neuronal growth and glial cells. AMP, adenosine monophosphate; ENS, enteric nervous system.

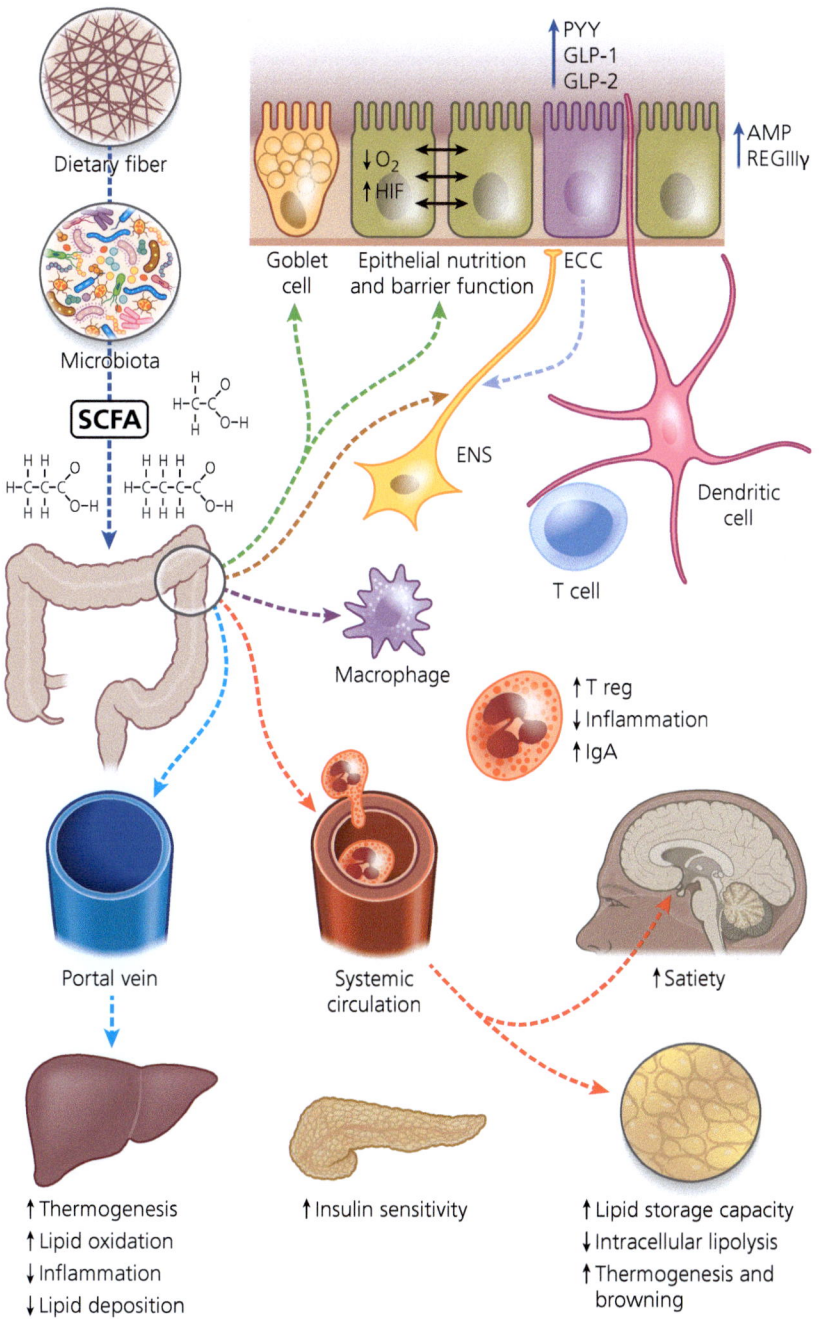

TABLE 5.1

Examples of non-fiber diet-microbe-host interactions

Dietary constituent	Microbial metabolism	Effect on host
Dietary fat	Increased generation of LPS; increased mucosal permeability and translocation of flagellin and LPS	Associated with inflammatory and metabolic disorders (fatty liver and insulin resistance)
Tryptophan	Three pathways: 1. Conversion by microbes to indoles – ligands for AHR 2. May also be metabolized by IDO1 to kynurenine pathway by immune and epithelial cells 3. Converted to serotonin by tryptophan hydroxylase in ECCs Pathways 2 and 3 are modified by the microbiota	Promotes production of anti-inflammatory IL-22 Impaired metabolism of tryptophan to AHR ligands is a risk factor for IBD, e.g. CARD9 deficiency Modulates signaling to the enteric and central nervous systems
Choline, glycine, betaine, L-carnitine	Production of TMA	TMA undergoes hepatic oxidation to TMA N-oxide which is associated with atherosclerosis
Histidine	Degraded by microbiota to imidazole proprionate	Imidazole proprionate is linked with insulin receptor degradation, also associated with diabetes and atherosclerosis
Oxalate	Degraded by *Oxalobacter formigenes* (present in about one third of individuals) and some lactobacilli	Altered risk of kidney stones

AHR, aryl hydrocarbon receptor; *CARD9*, caspase recruitment domain family member 9 [gene]; ECC, enterochromaffin cell; IBD, inflammatory bowel disease; IDO, indoleamine 2,3-dioxygenase; IL, interleukin; LPL, lipopolysaccharide; TMA, trimethylamine.

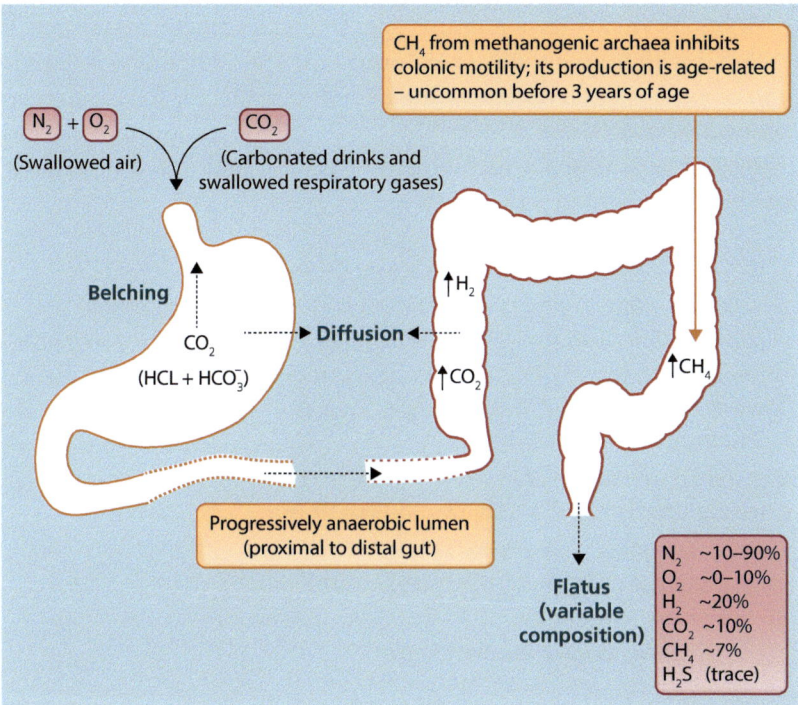

Figure 5.2 Gas production by microbes: gut microbiota produce many volatile organic compounds during metabolism of dietary constituents, but the main gases produced by microbes are hydrogen (H_2), carbon dioxide (CO_2) and methane (CH_4) with trace amounts of hydrogen sulfide (H_2S) in some individuals produced during the fermentation of proteins and by sulfur-reducing bacteria that reduce sulfates and sulfites. Nitric oxide (NO) may also be produced by intestinal bacteria by anaerobic denitrification. While oxygen (O_2) is a critical determinant of microbiota composition in the gut, it is derived almost exclusively from swallowed air and diminishes in concentration stepwise oro-anally. Likewise, nitrogen (N_2) is mainly derived from swallowed air (although tiny amounts may be produced by microbial denitrification of proteins in the colon); it is not metabolized and therefore its levels remain relatively stable during transit through the gut. Small amounts of gas (especially CO_2) may diffuse across the mucosa to and from the lumen and mucosal microvasculature.

preliminary results have prompted controlled feeding interventions in children with acute moderate malnutrition. Thus, microbial responses and blood protein levels following microbiota-directed foods have been superior to those achieved by standard food interventions.[10,11] The usual caveats apply, including the need for replication in different ethnic groups and in larger studies.

The overnourished microbiota and obesity

Several lines of evidence point toward a role for the microbiota in obesity.[2,8] Biological plausibility is implied by the contribution of the microbiota to dietary energy harvest (germ-free [GF] animals consume up to 30% more calories than non-GF controls).

The most compelling experimental evidence is derived from the transfer of metabolic phenotypes including weight gain by fecal microbial transplants (FMT) from obese humans to GF animals. Similar human-to-human FMTs have so far been less impressive with only transient changes in the recipient's metabolic status. Moreover, no specific microbial signature has been consistently identified in obese individuals in different studies.

However, there is persuasive epidemiologic evidence suggesting that disruption of the assembly of the microbiome of children in the first year of life, particularly by antibiotics, may have contributed to the obesity epidemic in industrialized countries in combination with an obesogenic environment (widespread availability of energy-dense foods and reduced physical activity).[12] Low doses of antibiotics have been used to fatten animals, a high-fat diet does the same, and together they have an additive effect. Because of vertical transmission of the microbiome, antibiotic-induced loss of ancestral microbes may be cumulative down the generations. Human exposure to antibiotics is not restricted to prescribed medications; other potential routes of exposure include trace contamination of the food chain and water supply.

Processed foods

Processed foods are commonly claimed to have adverse effects on the microbiota and host health, but the evidence is limited and terms like 'processed' and 'ultra-processed' foods lack a uniformly-agreed definition. Although results in experimental animals seem to show that certain processed foods increase barrier permeability, the relevance to humans is unclear. The act of processing food per se is not necessarily harmful, but if processing removes fiber it can be construed as unhealthy when processed foods are the predominant habitual diet.

On the other hand, some but not all chemicals added during the processing of food may undergo microbial metabolism or may modify the microbiota with potentially unwanted consequences (Table 5.2).[13-18] It is noteworthy that many food additives have been introduced relatively recently and although they have GRAS status ('generally regarded as safe'), they may be metabolized by gut microbes and thus account for some of the features of the microbiomes of people living in industrialized countries.

Cooking as processing

Finally, cooking is an ancient form of food processing and increases the digestibility of starch thereby reducing amounts that reach the colonic microbiota. Cooking particularly affects the gut microbiomes of those consuming diets rich in tubers rather than high meat diets.

Specialized diets

Differences in diet account for some but not all of the different microbiomes associated with different ethnic groups (see Figure 3.3, page 37). Specialized diets may also have distinctly different effects on the microbiota (Table 5.3).[19,20]

TABLE 5.2

Examples of food additives that alter the microbiota[13–18]

Additive	Use in food industry	Microbial response
Emulsifiers	Detergent-like agents to keep particles in suspension	Disrupt gastrointestinal mucus and alter composition of microbiota leading to inflammation, weight gain and metabolic syndrome in rodents
Artificial sweeteners	Soft drinks	Adverse metabolic impact on microbiota with glucose intolerance in the host
Trehalose	A stable disaccharide implemented in many foods over the past 2 decades	May have contributed to the emergence of epidemic strains of *C. difficile* infection. Certain ribotypes of *C. difficile* can metabolize trehalose and undergo increased virulence
Xanthan gum	An exopolysaccharide of *Xanthamonas campestris* increasingly used as a substitute for gluten	Capacity to digest xanthan gum is dependent on a single member of the Ruminococcaceae family of bacteria. Widespread use may contribute to some of the abrupt microbiome changes in industrialized countries

TABLE 5.3

Examples of specialized diet-microbiota interactions

Type of diet	Impact on microbiota and host
Restrictive or monotonous diets	Loss of dietary diversity leads to loss of gastrointestinal microbial diversity, which correlates with increased frailty in the elderly
Mediterranean diet*	Improves microbial diversity, slows age-related deterioration in microbiome composition, and reduces rate of progression of age-related frailty and cognitive function
Very low calorie diet (typically ~800 kcal/day)	Impact similar to that of parenteral nutrition Improved metabolic health but severe calorie restriction leads to reduced bacterial abundance and restructuring of the gut microbiome Weight loss is transmissible to mice colonized with post-diet microbiota
Low FODMAP (Fermentable Oligo-, Di-, Monosaccharides, and Polyols) diet	Loss of diversity and certain taxa such as bifidobacteria – dependent on duration of diet
Fermented foods (see ISAPP consensus reference[†])	Health benefits beyond nutritional value of ingredients because of variable content of viable microbes, SCFA, and bioactives such as bioactive peptides, bacteriocins and conjugated linoleic acid

CONTINUED

TABLE 5.3 CONTINUED

Examples of specialized diet-microbiota interactions

Type of diet	Impact on microbiota and host
Low protein diets for renal failure	Accumulation of gut microbial metabolites, such as p-cresyl sulfate, indole derivatives and trimethylamine N-oxide occurs in kidney failure, and reducing dietary protein and sulfur-containing amino acid intake is part of the management of chronic kidney disease. However, it has also been shown that dietary changes may also trigger posttranslational modification (S-sulfhydration) of microbial enzymes that reduce uremic toxin production, without altering the composition of the microbiota, in a murine model of chronic kidney disease

*Lower consumption of red meat, dairy products and saturated fats and increased intake of vegetables, fruits, legumes, fish, olive oil, and nuts.[21]
† Salminen S, Collado MC, Endo A, et al. The International Scientific Association of Probiotics and Prebiotics (ISAPP) consensus statement on the definition and scope of postbiotics. *Nat Rev Gastroenterol Hepatol.* 2021;18:649–667.
SCFA, short-chain fatty acids.

Key points – food and microbes

- Loss of dietary diversity leads to reduced microbial diversity.
- Microbes are net contributors to host nutrition and provide the metabolic capacity for energy extraction from dietary fiber.
- The end products of microbial fermentation of fiber include SCFA which greatly expand the functional impact of dietary fiber on intestinal and extraintestinal host physiology.
- Malnutrition in early life is associated with impaired host development and an immature microbiota. This is not resolved unless the provision of food also nourishes the microbiome.
- The contribution of the microbiota to dietary energy harvest has been implicated in obesity, but no specific microbial signature has been consistently linked with obesity.
- Persuasive experimental and epidemiologic evidence has linked obesity with a disturbed assembly of the microbiome in early life, particularly by exposure to antibiotics.

References

1. Su Q, Liu Q. Factors affecting gut microbiome in daily diet. *Front Nutr.* 2021; 8:644138.
2. Shanahan F, van Sinderen D, O'Toole PW, Stanton C. Feeding the microbiota: transducer of nutrient signals for the host. *Gut.* 2017;66:1709-1717.
3. Armet AM, Deehan EC, O'Sullivan AF, et al. Rethinking healthy eating in light of the gut microbiome. *Cell Host Microbe.* 2022;30:764-785.
4. Jardon KM, Canfora EE, Goossens GH, Blaak EE. Dietary macronutrients and the gut microbiome: a precision nutrition approach to improve cardiometabolic health. *Gut.* 2022;71:1214-1226.
5. Gill SK, Rossi M, Bajka B, Whelan K. Dietary fibre in gastrointestinal health and disease. *Nature Rev Gastroenterol Hepatol.* 2021;18: 101-116.

6. Deehan EC, Yang C, Perez-Muñoz ME, et al. Precision microbiome modulation with discrete dietary fiber structures directs short-chain fatty acid production. *Cell Host Microbe*. 2020;27:389-404.
7. Hughes RL, Kable ME, Marco M, Keim NL. The role of the gut microbiome in predicting response to diet and the development of precision nutrition models. Part II: results. *Adv Nutrition*. 2019;10:979-998.
8. Aron-Wisnewsky J, Warmbrunn MV, Nieudorp M, Klément K. Metabolism and metabolic disorders and the microbiome: the intestinal microbiota associated with obesity, lipid metabolism and metabolic health – pathophysiology and therapeutic strategies. *Gastroenterology*. 2021;160:573-599.
9. Robertson RC, Edens TJ, Carr L, et al. The gut microbiome and early-life growth in a population with high prevalence of stunting. *Nat Commun*. 2023;14:654.
10. Chen RY, Mostafa I, Hibberd MC, et al. A microbiota-directed food intervention for undernourished children. *N Engl J Med*. 2021;384: 1517-1528.
11. Chen RY, Mostafa I, Hibberd MC, et al. Melding microbiome and nutritional science with early child development. *Nat Med*. 2021;27:1503-1506.
12. Cox LM, Blaser MJ. Antibiotics in early life and obesity. *Nat Rev Endocrinol*. 2015; 11: 182–190.
13. Daniel N, Gewirtz AT, Chassaing B. Akkermansia muciniphila counteracts the deleterious effects of dietary emulsifiers on microbiota and host metabolism. *Gut*. 2023;72:906-917.
14. Ruíz-Ojeda FJ, Plaza-Díaz J, Sáez-Lara MJ, Gill A. Effects of sweetners on the gut microbiota: A review of experimental studies and clinical trials. *Adv Nutr*. 2019;10:S31-S48.
15. Nobs SP, Elinav E. Nonnutritive sweeteners and glucose intolerance: where do we go from here? *J Clin Invest*. 2023;133:e171057.
16. Yang G, Cao JM, Cui HL, Zhan XM, Duan G, Zhu YG. Artificial sweetener enhances the spread of antibiotic resistance genes during anaerobic digestion. *Environ Sci Technol*. 2023;57:10919-10928.
17. Collins J, Robinson C, Danhof H, et al. Dietary trehalose enhances virulence of epidemic Clostridium difficile. *Nature*. 2018; 553: 291-294.
18. Ostrowski MP, La Rosa SL, Kunath B, et al. Mechanistic insights into consumption of the food additive xanthan gum by the human gut microbiota. *Nat Microbiol*. 2022;7:556-569.

19. von Schwartzenberg RJ, Bisanz JE, Lyalina S, et al. Caloric restriction disrupts the microbiota and colonization resistance. *Nature*. 2021;595(7866):272-277.
20. Wastyk HC, Fragiadakis GK, Perelman D, et al. Gut-microbiota-targeted diets modulate human immune status. *Cell*. 2021;184: 4137-4153.
21. Ghosh TS, Rampelli S, Jeffery IB, et al. Mediterranean diet intervention alters the gut microbiome in older people reducing frailty and improving health status: the NU-AGE 1-year dietary intervention across five European countries. *Gut*. 2020;69(7):1218-1228.

6 Mindful microbes: brain–gut signaling

HEALTHCARE

Brain–gut axis

The transit of dietary intake from mouth to anus involves large volumes of gastrointestinal secretions with continual sampling, processing and selection of contents for absorption or excretion. For this to occur largely below the level of human consciousness, exquisitely precise coordination is required. This is accomplished in part by the brain–gut axis which is a communication system linking the central nervous system with the enteric nervous system so that the brain continually receives signals from the gut informing it of the luminal contents (including the microbiota) and their progression along the alimentary tract. The system is bi-directional with signals from the emotional and cognitive centers in the brain being delivered to the gut.[1-3]

The links between the brain and gut are both 'hard-wired' (involving the vagus nerve and spinal pathways) and indirect via the systemic circulation (involving an internet of crosstalk among the mucosal immune system, the neuroendocrine system and the nervous system with multiple feedback loops)[1,4] (Figure 6.1).

Figure 6.1 A brain–gut signaling internet. The gut mucosa is adapted to continual sampling of luminal contents, including microbial and dietary metabolites, primarily by M cells and dendritic cells, but with a variable degree of transepithelial uptake. These signals are sensed by the immune, endocrine, and nervous systems. Most of these metabolites act locally on the enteric nervous system either directly or via actions on enterochromaffin cells and immune cells including mast cells adjacent to mucosal neurons. The direct pathways to the central nervous system are the vagus nerve and spinal pathways but some metabolites act locally or are relayed centrally via the circulation. In response, the brain signals to the periphery by autonomic efferents also within the vagus and spinal pathways and via the hypothalamic-pituitary-adrenal axis. For example, efferent signals from the vagus pathway include regulatory effects on gastrointestinal motility and transit and an anti-inflammatory influence on the mucosal immune system. The schematic is intended to be representative not comprehensive. ACTH, adrenocorticotrophic hormone; CRH, corticotrophin-releasing hormone; ECC, enterochromaffin cell; ENS, enteric nervous system; LPS, lipopolysaccharide; SCFA, short-chain fatty acids.

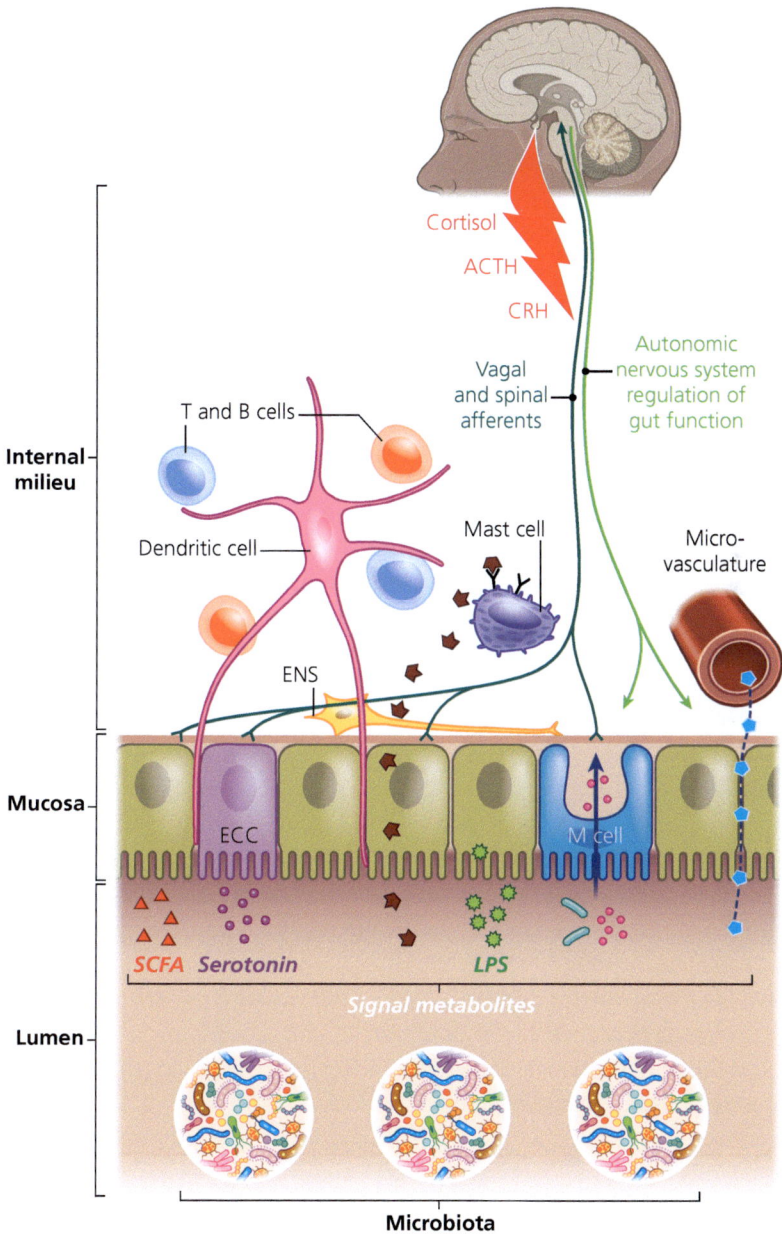

Evidence for microbe-to-brain signaling

Several lines of evidence for the influence of microbial signals on brain structure and function have been reported. Firstly, hepatic encephalopathy is caused by the toxic effects of gut-derived microbial metabolites which are normally metabolized by the liver but which escape into the systemic circulation if the liver is diseased or bypassed. The condition is alleviated effectively by targeting the microbiota with topical antibiotics or laxatives. Secondly, the influence of the microbiota on brain development and behavior is shown by comparative studies of germ-free (GF) animals and those which have been conventionally raised (as discussed in Chapter 2.).[5] Thirdly, disturbances of the microbiota due to antibiotics particularly in early life may have an enduring adverse influence on the central nervous system. Fourthly, several preclinical studies with experimental animals, usually mice, have been reported to show a modifying influence of the microbiota on brain behavior, but the degree to which such studies can be extrapolated to humans is unclear.[1,2,6] Finally, interventions in humans with putative probiotics or food-grade organisms may signal from gut to brain but the durability of the responses is uncertain and results from different studies have been inconsistent or not replicated.[7-9]

Signaling molecules

Specific microbes have been shown to produce neurotransmitters in vitro such as gamma-aminobutyric acid and dopamine but it is uncertain if these are produced in vivo in sufficient concentrations to have local effects and they are unlikely to penetrate the blood-brain barrier.[1,2]

Three main categories of microbiota-derived signaling molecules have been identified which may have local or systemic actions.

- Metabolites of dietary polysaccharides, amino acids and polyphenols which may have local or systemic effects once released by microbial metabolism. Examples include short-chain fatty acids (SCFA) generated by gut microbes by fermentation of dietary fiber which exert receptor-mediated actions on various cell types including neuronal, enterochromaffin, immune and other cells and which regulate satiety and immune activation.[10] In addition, microbes metabolize dietary tryptophan and produce the neuroactive metabolites 5-hydroxytryptamine (serotonin), kynurenine and

indoles. Serotonin may act locally on adjacent cells including neurons or act in an endocrine fashion to influence the development and function of the enteric and central nervous systems.[11]
- Microbial cell wall components such as lipopolysaccharide (LPS) and peptidoglycans which act on Toll-like receptors (TLRs) on neurons, microglia and immune cells with release of cytokines and downstream effects.[11,12]
- Metabolites derived from microbial action on host-derived molecules such as bile acids and sex hormones.[1,2]

Putative disorders of the gut-brain axis

Irritable bowel syndrome (IBS) which is regarded as the prototype disorder of the gut-brain axis, is a chronic visceral pain disorder, the diagnosis of which is based on typical symptoms rather than signs or laboratory abnormalities and on the exclusion of other disorders.[7,13,14] It affects 5–10% of the population in most countries and is far more common in premenopausal women than men.

Although IBS is clinically heterogeneous, hypersensitivity of visceral pain sensation is evident in most patients. Metabolites such as histamine produced by some microbes may have a contributory role.[15] Recent findings also implicate a role for enterochromaffin (EC) cells in the gut epithelium which sense dietary, microbial and other factors in the lumen and activate adjacent enteric nerves using a serotonin-dependent mechanism.[16] By genetic and pharmacological manipulation of the EC cell-sensory nerve circuit in mice, EC cells were shown to be the main drivers of visceral pain. Moreover, continual activation of this circuit leads to persistence of visceral hypersensitivity with anxiety-like behavior and striking sex differences, thereby replicating key features of human IBS.

A direct causal role for the microbiota in IBS has not been established. Studies of IBS have yielded inconsistent results which reflect differences in study design but also the heterogeneity of IBS and its comorbidities.[7] Similarly, microbiota-based therapies such as fecal microbial transplants (FMT) have been inconsistent, although specific bifidobacterial strains but not all putative probiotics appear to be effective.[7]

Although the microbiota-gut-brain axis has been implicated in other disorders such as Parkinson's disease and autism, the evidence is inconclusive and based largely on animal models.[17]

Key points – mindful microbes: gut-brain signaling

- The brain–gut axis a bi-directional communication system that ensures that the brain receives and responds to signals from the gut informing it of the luminal contents including the microbiota.
- The links between the brain and gut are both hard-wired (vagus nerve and spinal pathways) and indirect via the systemic circulation.
- Signaling molecules include dietary metabolites such as SCFA and serotonin and microbial cell wall components.
- The influence of the microbiota on brain development and behavior has been shown by comparative studies of germ-free animals and those which have been conventionally raised.
- The degree to which the microbiota influences human behavior is uncertain.
- The microbiota-gut-brain axis has been implicated in IBS and a miscellany of disorders including Parkinson's disease and autism but the evidence is inconclusive and is based largely on animal studies.

References

1. Mayer EA, Nance K, Chen S. The gut-brain axis. *Annu Rev Med*. 2022;73:439–453.
2. Margolis KG, Cryan JF, Mayer EA. The microbiota-gut-brain axis: from motility to mood. *Gastroenterology*. 2021;160:1486–1501.
3. Gershon MD, Margolis KG. The gut, its microbiome, and the brain: connections and communications. *J Clin Invest*. 2021;131(18):e143768.
4. Morais LH, Schreiber HL, Mazmanian SK. The gut microbiota-brain axis in behaviour and brain disorders. *Nat Rev Microbiol*. 2021;19:241–255.
5. Luczynski P, McVey Neufeld KA, Oriach CS, Clarke G, Dinan TG, Cryan JF. Growing up in a bubble: using germ-free animals to assess the influence of the gut microbiota on brain and behavior. *Int J Neuropsychopharmacol*. 2016;19(8):pyw020.

6. Vanuytsel T, Bercik P, Boeckxstaens G. Understanding neuroimmune interactions in disorders of gut-brain interaction: from functional to immune-mediated disorders. *Gut*. 2023;72:787–798.
7. Shanahan F, Quigley EM. Manipulation of the microbiota for treatment of IBS and IBD- challenges and controversies. *Gastroenterology*. 2014;146:1554–1563.
8. Martin CR, Osadchiy V, Kalani A, Mayer EA. The brain-gut-microbiome axis. *Cell Mol Gastroenterol Hepatol*. 2018;6:133–148.
9. Berding K, Cryan JF. Microbiota-targeted interventions for mental health. *Curr Opin Psychiatry*. 2022;35:3–9.
10. O'Riordan KJ, Collins MK, Moloney GM, et al. Short-chain fatty acids: microbial metabolites for gut-brain axis signalling. *Mol Cell Endocrinol*. 2022;546:111572.
11. Cryan JF, O'Riordan KJ, Cowan CSM, et al. The microbiota-gut-brain axis. *Physiol Rev*. 2019;99:1877–2013.
12. Ratsika A, Cruz Pereira JS, Lynch CMK, Clarke G, Cryan JF. Microbiota-immune-brain interactions: a lifespan perspective. *Curr Opin Neurobiol*. 2023;78:102652.
13. Niesler B, Kuerten S, Demir IE, Schäfer K-H. Disorders of the enteric nervous system – a holistic view. *Nat Rev Gastroenterol Hepatol*. 2021;18:393–410.
14. Bercik P. The brain-gut-microbiome axis and irritable bowel syndrome. *Gastroenterol Hepatol*. 2020;16(6):322–324.
15. De Palma G, Shimbori C, Reed DE, et al. Histamine production by the gut microbiota induces visceral hyperalgesia through histamine 4 receptor signaling in mice. *Sci Transl Med*. 2022;14(655):eabj1895.
16. Bayrer JR, Castro J, Venkataraman A, et al. Gut enterochromaffin cells drive visceral pain and anxiety. *Nature*. 2023;616(7955):137–142.
17. Claudino Dos Santos JC, Oliveira LF, Noleto FM, Gusmão CTP, Brito GAC, Viana GSB. Gut-microbiome-brain axis: the crosstalk between the vagus nerve, alpha-synuclein and the brain in Parkinson's disease. *Neural Regen Res*.2023;18(12):2611–2614.

Gastroenterology

7 Microbes and chronic intestinal disease

HEALTHCARE

Although the gut microbiome has a proven causative role in a relatively small number of human diseases (see Table 3.1, page 36), there is persuasive evidence linking the rising prevalence of chronic non-communicable diseases (NCCDs) with the changing nature of the microbiome and modern lifestyles in socioeconomically developed countries. In contrast to the human genome which adapts slowly, the microbiome evolves at a greater pace in response to environmental changes linked with socioeconomic development and industrialization. Moreover, the striking increases in incidence and prevalence in Western societies of NCCDs, such as Crohn's disease and ulcerative colitis, have occurred over too short an interval to be due to changes in genetic risk factors.

Missing microbes and migrant microbes

The original conception of the 'hygiene hypothesis' mistakenly attributed the cause of many chronic diseases to excessive cleanliness and reduced exposure to environmental pathogens. A more likely explanation is the reduced exposure to protective organisms associated with modern lifestyles. Modernization and socioeconomic development are linked with reduced microbial diversity, loss of ancestral organisms, such as *Helicobacter pylori* and *Oxalobacter*, and a decline of endemic parasitism.

The influence of lifestyle and the environment on the microbiota is greatest during the earliest stages of human life, when the microbiota is being assembled and while the immune and metabolic pathways are simultaneously maturing.[1] Although the immune system is fully formed in a full-term neonate, it still requires educational input from the microbial environment to have the capacity to distinguish danger from non-danger. Thus, microbial education of the immune system is achieved by exposure to a diversity of harmless microbes and microbial metabolites during colonization in the first few years of life. When this process is delayed or disrupted, such as by exposure to antibiotics, there is an increased risk of disturbed immune perception of self and non-self with the potential for autoimmune and allergic disorders in adolescence or early adulthood.

Migration studies support the particular importance of lifestyle and environmental influences on the microbiota in early life and the consequential risk of developing a chronic disease. For example, the earlier a migrant relocates from a socioeconomically undeveloped region of low-disease-prevalence to an industrialized country with a high prevalence of chronic diseases, the greater the risk of developing inflammatory bowel disease.[1] In contrast, the risk remains low if migration occurs later in life.

As expected, the microbiome of migrants to modern industrialized countries changes commensurate with the duration of exposure to their new world environment. Progressive westernization of the microbiome has been observed in migrants to the United States, with changes in microbial diversity, strain composition and function. Changes in diet account for much, but not all, of the westernization of the microbiome of migrants. The westernization also becomes more evident in successive generations of immigrant families, consistent with the view that modernization leads to loss of protective components of the microbiota.

Inflammatory bowel disease (Crohn's disease and ulcerative colitis)

Background and heterogeneity. Inflammatory bowel disease (IBD) encompasses Crohn's disease and ulcerative colitis. However, this collective term obscures how distinct and heterogeneous each of these disorders is. In contrast to Crohn's disease which may affect any part of the digestive tract from mouth to anus and penetrate the gut wall, ulcerative colitis is confined to the large intestine (colon and rectum) and usually limited to the mucosa. Moreover, ulceration is the early defining feature of Crohn's disease, whereas ulceration is uncommon or appears late in severe–not mild–ulcerative colitis. Differences in immune reactivity to the microbiota are exemplified by serologic antibodies to microbial antigens that cross react with neutrophils in about 60% of patients with ulcerative colitis, whereas in a subset (~40%) of patients with Crohn's disease there are antibodies to *Sacharomyces cerevisiae* and T-cell reactivity to multiple Lachnospiraceae flagellins.[2]

The heterogeneity of both conditions is also reflected in the wide range of different genetic defects of mucosal homeostasis in different experimental animal models, which may lead to similar pathology (Box 7.1).

> **BOX 7.1**
>
> **Lessons from animal models of IBD**
>
> - Multiple pathways and distinct genetic defects may lead to similar disease phenotypes
> - Disease susceptibility may be at the level of the immune system of end organ
> - Prevention of inflammation requires active regulation of mucosal homeostasis
> - Gene-gene and gene-environment interactions modify severity and phenotype
> - The microbiota is required for pathogenesis but is not necessarily causative*
> - Treatment efficacy in animal models often fails to translate to the human condition
> - Most models are monophasic, not polyphasic (relapsing and remitting) as in humans
>
> *The microbiota is required for immune development to be sufficient to mount an inflammatory response. For example, in a highly informative model of murine colitis, involving a genetic predisposition, the microbiota was necessary but not sufficient to cause colitis triggered by a norovirus and/or chemical exposure.[3]

Studies of identical and non-identical twins show that genetic risk is far more pronounced in Crohn's disease than in ulcerative colitis. Over 200 genetic risk alleles for both conditions have been identified, many of which are shared, and some are distinct to either Crohn's or colitis.[4] The genetic risk factors code mainly for proteins involved in immune regulation, the host immune response to microbes and mucosal barrier function. These genetic risk factors are common in society and are insufficient alone to cause disease in most instances.

Environmental or lifestyle risk factors, chief of which is socioeconomic development, are also common to both Crohn's disease and ulcerative colitis, but a noteworthy exception is tobacco smoking which is more common in Crohn's disease and exacerbates the condition, whereas the cessation of smoking often precedes or aggravates ulcerative colitis. Smoking is known to modify the microbiome, but it is unclear if this contributes to its opposing influence on the two forms of IBD.

The microbiota in Crohn's disease and ulcerative colitis

In view of the clinical and genetic heterogeneity of Crohn's disease and ulcerative colitis, it is unlikely that a single microbial configuration accounts for all forms of the disease.[5] In a systematic review of studies of patients with IBD, evidence for differences in the abundance of some bacteria was found in patients compared with controls, but results and methods across different studies were inconsistent.[6] Reduced microbial diversity has been observed in almost all studies but this is a non-specific finding found in almost all acute illnesses. It may be a stress response to illness or a result of dietary changes consequent to illness.

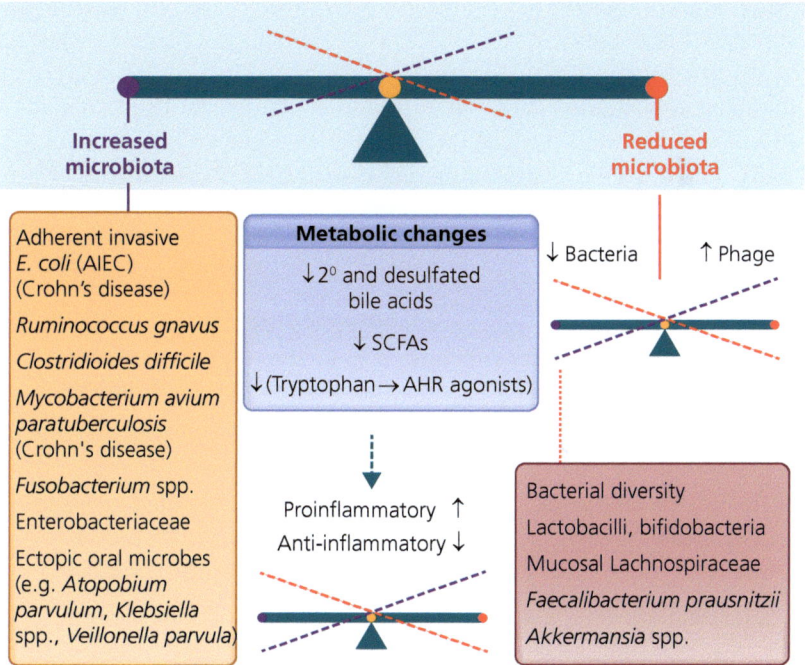

Figure 7.1 Gut microbiota identified to be increased or decreased in abundance in most, but not all, studies of patients with active inflammatory bowel disease. Collectively, these microbial disturbances are associated with a shift toward a pro-inflammatory response and alterations in bile acid metabolism, tryptophan metabolism and SCFA production. AHR, aryl hydrocarbon receptor; SCFA, short-chain fatty acid.

Discussions of the microbiota in IBD are usually cast as an imbalance with loss of protective commensals and expansion of pathobionts including a secondary expansion of Proteobacteria due to changes in oxygen tension in the inflamed bowel (Figure 7.1). The microbial imbalance is usually linked with a corresponding imbalance in anti-inflammatory and pro-inflammatory microbial metabolites. This binary language and logic may be simplistic and does not take into account interactions among microbes. Moreover, cross-sectional studies in humans with Crohn's and ulcerative colitis show that microbial disturbances are strongly associated with active inflammation and are less evident in areas of the bowel that are non-inflamed.[7] In addition, longitudinal studies of the same patients in relapse and remission are consistent with many of the microbial abnormalities being secondary to the inflammatory state.[8] However, there is evidence for different microbes contributing to different phases or complications particularly in advanced Crohn's disease[9-14] (Table 7.1). In addition, prospective study of first-degree relatives of patients with Crohn's disease has found a preclinical gut microbiome signature associated with future development of Crohn's disease

TABLE 7.1

Examples of microbial contribution to different phases, aspects or complications of established inflammatory bowel disease[9-14]

Stage or complication of disease	Contributory microbial organism(s)	Comment
Co-infection (enhanced inflammation)	*C. difficile*, CMV, *Candida albicans*	Crohn's disease and ulcerative colitis
Abscess formation	*Bacteroidetes*	Crohn's disease
Creeping fat	*Clostridium innocuum*	Crohn's disease
Fibrosis/strictures	Anaerobes including *C. ramosum, B. fragilis, B. uniformis*	Crohn's disease
Delayed wound healing	*Debaryomyces hansenii*	Crohn's disease
Colitis-associated cancer	*Morganella morganii, E. coli, Clostridium perfringens*	Strain-specific production of genotoxic metabolites

Regardless of cause or consequence, the altered microbiota associated with active IBD seems to amplify the inflammatory process, in part by disturbing several metabolic pathways including bile acid metabolism (reduced secondary bile acids, impaired desulfation, and increased primary bile acids), reduced short-chain fatty acid (SCFA) production and altered metabolism of dietary tryptophan with reduced generation of aryl hydrocarbon receptor (AHR) agonists such as indole-3-acetic acid – all of which tend to shift the mucosal milieu toward a pro-inflammatory state (see Figure 7.1, page 79).[11,14,15]

Although most studies have focused on the bacteriome, there is increasing evidence for altered phage and other virome signatures, including increased Caudovirales and a reduction in Microviridae in IBD. It is unclear at present if these changes are related to disturbances in the bacteriome and whether they are a cause or a consequence of disease.

The microbiota and other chronic intestinal disorders

Irritable bowel syndrome (IBS). In contrast to IBD, the diagnosis of IBS is not based on objective evidence of abnormality; it relies on the patient's symptoms. Studying IBS is also complicated by the lack of reliable animal models. A role for the microbiota in IBS is less well substantiated than in IBD.[9] Circumstantial evidence includes:
- the occurrence of a postinfectious form of IBS which accounts for a small subset of patients
- a controversial weak association between *H. pylori* infection and functional dyspepsia (considered to be a form of IBS)
- a reported link between antibiotic exposure and risk of developing IBS
- therapeutic responsiveness to some but not all probiotics in some patients.

However, several studies have shown alterations in the fecal microbiota of patients with IBS; the changes are subtle and inconsistent across different study populations. While some reports show a correlation with symptom severity, there is no clear relationship between microbial disturbances and the clinical subtypes of IBS (diarrhea-predominant, constipation-predominant and mixed/alternating).[16] It is also unclear whether the microbiota disturbances in IBS are secondary to the condition, its treatment, including dietary adjustments, or due to psychological comorbidity.

A more direct relationship between the microbiota and IBS has been proposed for an apparent subset of patients with small intestinal bacterial overgrowth (SIBO).[17] Classic cases of SIBO were recognized by association with severe anatomic or functional abnormalities of the intestine, but more recently a more subtle form of SIBO has been proposed to account for the symptoms of some patients with IBS (Box 7.2).

BOX 7.2

Small intestinal bacterial overgrowth (SIBO)

- Definitions of SIBO are vague and refer to increased bacterial numbers in the small bowel without a uniformly agreed normal range
- Historically SIBO was named 'blind loop syndrome' when clinicians found an objectively defined anatomic abnormality and bacterial counts of $>10^5$ CFU/ml cultured from a jejunal aspirate
- Predisposing causes include stasis or intestinal hypomotility including diabetic autonomic neuropathy, scleroderma, jejunal diverticulosis, intestinal pseudo-obstruction and surgically created blind loops or other anatomic anomalies. In addition, hypochlorhydria due to proton pump inhibitors (PPIs) may predispose to SIBO although evidence is inconclusive. Similarly, SIBO may occur in people with hypogammaglobulinemia
- Disease mechanisms: direct mucosal injury, competition with the host for nutrients, and pathophysiological effects of bacterial metabolites
- A diversity of clinical presentations including IBS has been attributed to SIBO based on poorly validated clinical tests but without any objectively defined structural or functional abnormality
- SIBO is long established as a cause of malabsorption and diarrhea, but evidence for its contribution to symptoms of IBS is controversial and unproven. This will only be resolved by an accurate definition and a standard metric for the normal small intestinal microbiome

---

CONTINUED

> CONTINUED
> - Currently, the likelihood of a positive test for SIBO varies according to the methodology used (culture vs breath test) and the geographic residence of the patient
> - SIBO is more common in the elderly
> - Breath tests for SIBO are based on the principle that the metabolism of carbohydrate substrates such as lactulose or glucose by microorganisms results in the production of hydrogen or methane that are then absorbed and exhaled as gas in the breath
> - The concordance between jejunal aspiration and breath test results is poor

Celiac disease. Although the essential elements for the pathogenesis of celiac disease are genetic susceptibility (risk alleles HLA-DQ2, HLA-DQ8, and HLA-DQ7) and exposure to dietary gluten, these are insufficient alone to cause the disease; other cofactors include potential viral infections and a role for the gut microbiota in proteolysis of gluten and in modifying the severity of tissue injury.[18] Variations in the small bowel microbiota exhibit high interindividual variation and also vary at different sampling sites.

Key points – microbes and chronic intestinal disease

- The increasing prevalence of NCCDs such as IBD has been linked with modern lifestyles and a changing microbiome in socioeconomically developed countries.
- Migration studies show that lifestyle and environmental influences on the microbiota are greatest in early life.
- Progressive reduction in microbial diversity and changes in composition have been observed following migration from developing to socioeconomically developed countries.
- Changes in diet account for much, but not all, of the westernization of the microbiome of migrants.
- Although changes in the microbiota associated with Crohn's disease and ulcerative colitis may not cause these diseases, there is persuasive evidence for different microbes contributing to different phases or complications, particularly in advanced Crohn's disease.
- Regardless of cause or consequence, the altered microbiota associated with active IBD seems to amplify the inflammatory process, in part by disturbing several metabolic pathways.

References

1. Shanahan F, Ghosh TS, O'Toole PW. The healthy microbiome – what is the definition of a healthy gut microbiome? *Gastroenterology*. 2021;160:483-494.
2. Alexander KL, Zhao Q, Reif M, et al. Human microbiota flagellins drive adaptive immune responses in Crohn's disease. *Gastroenterology*. 2021;161:522-535.
3. Cadwell K, Patel KK, Maloney NS, et al. Virus-plus-susceptibility gene interaction determines Crohn's disease gene Atg16L1 phenotypes in intestine. *Cell*. 2010;141:1135-1145.
4. Noble AJ, Nowak JK, Adams AT, et al. Defining interactions between the genome, epigenome, and the environment in inflammatory bowel disease: progress and prospects. *Gastroenterology*. 2023;165(1):44-60.e2.

5. Lee M, Chang EB. Inflammatory bowel diseases (IBD) and the microbiome – searching the crime scene for clues. *Gastroenterology.* 2021;160:524-537.
6. Pittayanon R, Lau JT, Leontiadis GI, et al. Differences in gut microbiota in patients with vs without inflammatory bowel diseases: a systematic review. *Gastroenterology.* 2020;158: 930-946.
7. Ryan FJ, Ahern AM, Fitzgerald RS, et al. Colonic microbiota is associated with inflammation and host epigenomic alterations in inflammatory bowel disease. *Nat Commun.* 2020;11:1512.
8. Clooney AG, Eckenberger J, Laserna-Mendieta E, et al. Ranking microbiome variance in inflammatory bowel disease: a large longitudinal intercontinental study. *Gut.* 2021;70:499-510.
9. Shanahan F, Quigley EMM. Manipulation of the microbiota for treatment of IBS and IBD – challenges and controversies. *Gastroenterology.* 2014;146:1554-1563.
10. Cao Y, Oh J, Xue M, et al. Commensal microbiota from patients with inflammatory bowel disease produce genotoxic metabolites. *Science.* 2022; 378:369.
11. Benech N, Sokol H. Targeting the gut microbiota in inflammatory bowel diseases: where are we? *Curr Opin Microbiol.* 2023;74:102319.
12. Ha CWY, Martin A, Sepich-Poore GD, et al. Translocation of viable gut microbiota to mesenteric adipose drives formation of creeping fat in humans. *Cell.* 2020;183:666-683.
13. Jain U, Ver Heul AM, Xiong S, et al. Debaryomyces is enriched in Crohn's disease intestinal tissue and impairs healing in mice. *Science.* 2021; 371:1154-1159.
14. Caruso R, Lo BC, Núñez G. Host–microbiota interactions in inflammatory bowel disease. *Nat Rev Immunol.* 2020;20:411-426.
15. Thomas JP, Modos D, Rushbrook SM, Powell N, Korcsmaros T. The emerging role of bile acids in the pathogenesis of inflammatory bowel disease. *Front Immunol.* 2022; 13:829525.
16. Jeffery IB, Das A, O'Herlihy E, et al. Differences in fecal microbiomes and metabolomes in people with and without irritable bowel syndrome and malabsorption. *Gastroenterology.* 2020;158(4):1016-1028.
17. Bushyhead D, Quigley EMM. Small intestinal bacterial overgrowth – pathophysiology and its implications for definition and management. *Gastroenterology.* 2022;163: 593–607.
18. Constante M, Libertucci J, Galipeau HJ, et al. Biogeographic variation and functional pathways of the gut microbiota in celiac disease. *Gastroenterology.* 2022;163:1351-1363.

8 Cancer and the microbiome

Gastroenterology

HEALTHCARE

Beyond cause and effect

Until recently, the link between microbes and cancer was addressed mainly in terms of cause and effect, but a microbial role in diagnosis or treatment of cancer was long disputed. While chronic infections still account for a substantial proportion of cancers (~13% of global cancer cases)[1] and may accelerate the progression of certain other cancers (Table 8.1; Figure 8.1),[1-4] mobilizing personal microbes in the fight against cancer has emerged as an exciting and realistic prospect.

TABLE 8.1

Examples of microbes implicated in cancer causation

Microbe	Cancer*
Helicobacter pylori	Adenocarcinoma of body of stomach**
	Lymphoma of stomach
Hepatitis B virus	Hepatoma
Hepatitis C virus	Hepatoma
Human immunodeficiency virus (HIV)	Kaposi's sarcoma
Epstein–Barr virus (EBV)	Burkitt's lymphoma
	Nasopharyngeal cancer
Human papilloma virus	Cervical cancer

*With cancers such as colorectal carcinoma, the microbe-cancer relationship may be more complex than one-microbe-one-disease, including tumor fungi that do not initiate carcinogenesis but may accelerate it or alter the clinical course of cancer.[4-9] Colorectal cancer development has been consistently linked with three organisms: (a) colibactin-producing *Escherichia coli* which causes neoplastic DNA damage; (b) enterotoxigenic *Bacteroides fragilis* which causes toxin-induced cell proliferation and tumor-promoting inflammation; and (c) *Fusobacterium nucleatum* which enhances colonic cancer progression mediated by the adhesins Fap2 and FadA, which drive proliferation and antitumor immune evasion.

**H. pylori* causes cancer of the body of the stomach but is protective against development of adenocarcinomas of the upper stomach and lower end of esophagus.[10]

The microbiome and host defense against cancer

Although many cancers find ways to evade or subvert the immune system, the microbiome has evolved clever strategies to defend against cancer particularly by influencing the host immune defense. Microbial metabolites and cell wall components prime an adaptive immune response against tumor cells by engaging with pattern recognition receptors on dendritic and other antigen-presenting cells. This leads to cytokine release and activation of immune effector cells which may act locally or systemically (Figure 8.2).

Immune stimulation by the gut microbiota is probably insufficient alone to eradicate most established cancers, but its contribution has become evident in the setting of immunotherapy, particularly with checkpoint inhibitors.[11,12] Immune checkpoints are naturally occurring proteins which prevent excessive immune responses. If checkpoint receptors on T cells bind to checkpoint proteins on cancer cells, the T cell is switched off, thereby inhibited from rejecting the cancer. Checkpoint inhibitory drugs prevent this by targeting either the checkpoint proteins such as programmed cell death protein 1 (PD-1) (or CTLA-4) or their partner protein ligands e.g. programmed death-ligand 1 (PD-L1) (Figure 8.2d).

Tumor-associated microbiota

While cancers of the colon and to a lesser extent cancers at other mucosal sites are exposed to luminal microbes, seeding of tumors in non-mucosal internal organs, such as with skin melanomas, is also common. This usually occurs by translocation from mucosal tissues and occasionally by direct infection. Intriguingly, translocation of gut microbes by dendritic cell uptake and transit via the mesenteric lymph nodes has been shown to be enhanced by therapy with immune checkpoint inhibitors.[12]

The influence of intratumoral bacteria on cancer outcome is variable. In some cases, intratumoral bacteria have been shown to alter the local concentrations of chemotherapeutic drugs by enzymatic inactivation. *Fusobacterium nucleatum* may also confer resistance to chemotherapy by activating Toll-like receptors (TLRs) on cancer cells and modulating autophagy. In addition, preclinical experimental studies have shown that the presence of bacteria within the tumor microenvironment may have varying immunostimulatory or immunosuppressive effects on the host immune response.[13] For example, profiling pancreatic tumor microbiota not only has prognostic utility but there also appears to be a role for antibiotic-mediated reduction of the intratumoral bacterial load in pancreatic cancer which reduces the recruitment of suppressor cells and favors recruitment of innate effector T cells.

Some bacteria may enter a cancer cell and shed protein fragments which are transported to the surface of the cancer cell where they are presented to the immune system (Figure 8.2b). This enables the immune system to distinguish the cancer cell as foreign and selectively attack without injuring the surrounding tissue.[14,15] Research into this process raises the prospect of coopting bacterial and other microbial proteins to enhance immune rejection and to refine the treatment of cancer by minimizing collateral damage. However, not all microbes are equally effective in promoting an antitumor response, and this is reflected in human-to-human variability in the capacity of the gut microbiome to influence the clinical course of cancer.

Influence of cancer therapy on the microbiota

Most aspects of cancer care affect the composition and function of the gut microbiota.[3] Chemotherapy may alter the microbiota and affect multiple metabolic pathways. Radiotherapy has a diversity of effects on host-microbe interactions, most notably disruption of the integrity of the mucosal barrier. Broad-spectrum antibiotics which are often required during chemotherapy, particularly for patients with leukopenia and for patients undergoing surgery for cancer, have profound effects on the microbiota. These have been shown to adversely affect the outcome of cancer immunotherapy.

Figure 8.1 Representative examples of microbiota-mediated mechanisms of carcinogenesis within the gastrointestinal tract and at extraintestinal sites. The relationship between microbes and susceptibility to cancer is complex and multifactorial. Note the striking difference in incidence of cancer across the ileocecal valve, i.e. between small and large bowel, even though these segments of the gastrointestinal tract are in direct continuity and exposed mainly to the same microbes.

Cancer and the microbiome

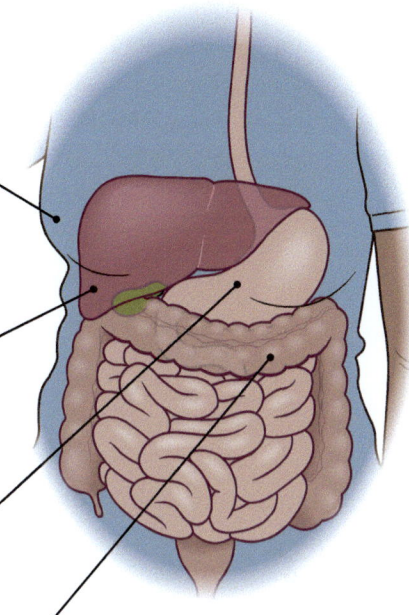

Breast
- Microbiota-mediated altered circulating estrogen levels
- Microbiota contribution to fat metabolism and obesity

Liver/biliary tract
- Gut microbial metabolites from portal circulation
- Generation of secondary bile acids may alter immune function and tumor growth
- Genotoxicity and hepatotoxicity induced by secondary bile acids
- Induction of non-alcoholic steatohepatitis that predisposes to cirrhosis and cancer

Stomach
- *H. pylori* induced chronic inflammation
- Genotoxic effects on gastric mucosa

Large bowel (colon and rectum)
- Chronic mucosal inflammation
- Microbial-mediated mucin degradation and barrier impairment
- Production of genotoxic metabolites
- Local generation of reactive oxygen species
- Induction of oncogenic transcriptional activity potentially via β-catenin–Wnt signaling pathways
- Altered antitumor immune defense

Small bowel
- Cancer risk – **low**
- Surface area +++
- Epithelial turnover ++
- Apoptosis +++
- Bacterial load +

Large bowel
- Cancer risk – **high**
- Surface area +
- Epithelial turnover +
- Apoptosis +/–
- Bacterial load +++

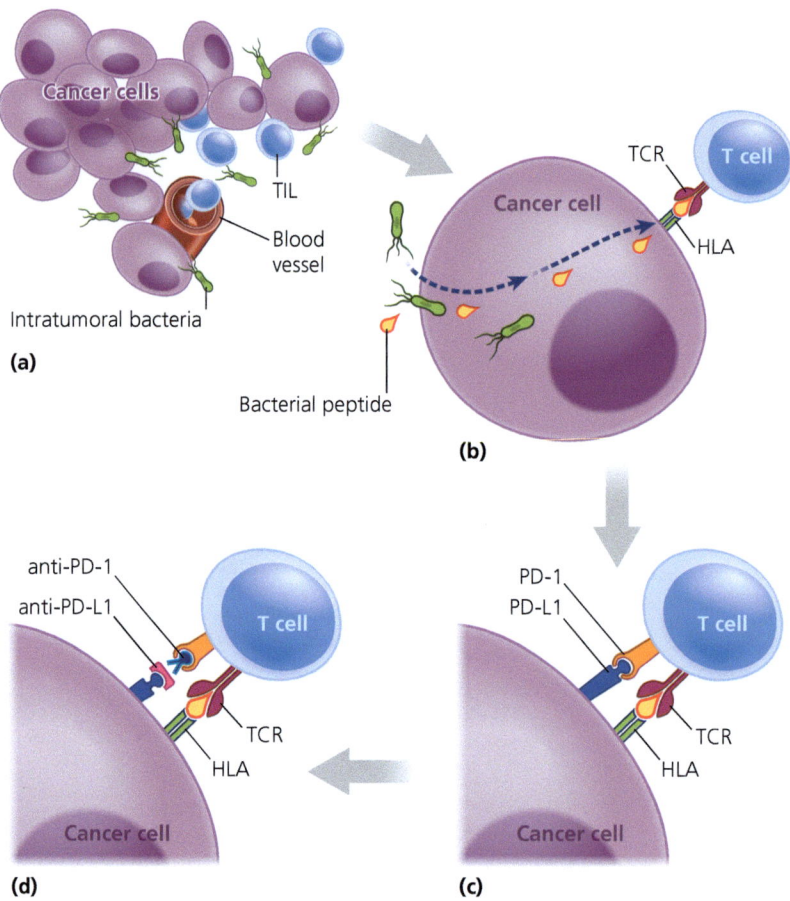

Figure 8.2 The systemic inflammatory tone of the immune system is shaped by the microbiota. (a) Bacteria commonly seed cancers including those remote from mucosal tissues where they engage with immune cells which generate cytokines and recruit tumor-infiltrating lymphocytes (TIL). (b) In some instances, bacteria gain entrance to cancer cells and shed proteins which are presented to T cells in the context of self (with human leukocyte antigens [HLA]). However, this is seldom sufficient to reject a well-established cancer but when patients are treated with immunotherapy using immune checkpoint inhibitors (c) and (d) the tumor microbial metabolites enhance the anti-cancer immune response. PD-1, programmed cell death protein 1; PD-L1, programmed death-ligand 1; TCR, T-cell receptor.

Influence of the microbiome on cancer chemotherapy

The gut microbiota influences not only the response, but also the toxicity of different types of cancer therapy, including chemotherapy, immune checkpoint inhibition, and stem cell transplantation (see Chapter 12 for comparison). The story of irinotecan is an intriguing example.[16] Irinotecan has been used in combination with other drugs to treat various cancers including colonic, ovarian and pancreatic cancer. The drug is administered intravenously and is inactivated in the liver by glucuronidation and excreted via the bile duct into the gastrointestinal tract. However, when it encounters gut bacteria it is reactivated by microbial β-glucuronidase which removes the glucuronide group. It is then toxic to the epithelial cells lining the gut, resulting in dose-limiting diarrhea that may be incapacitating. Inhibition of microbial β-glucuronidase alleviates this toxicity without killing gut bacteria.[16]

Translating microbiome science to cancer care

Diagnostic profiling. Advances in linking the microbiota with risk of disease raises the prospect of profiling of the gut microbiota and tumor microbiota for both diagnostic and prognostic purposes. The most promising data relate to colorectal cancer. Biomarkers that are currently available detect established disease early but microbiome profiling might soon be able to identify microbial risk factors for disease decades before the age at which current screening programs commence. Profiling the pancreatic tumor microbiota has been reported to be of prognostic value in separating long- from short-term survivors.[13]

Modifying the microbiota. Several studies have shown the value of modifying the microbiota to enhance immune responses against cancer, particularly when combined with immune checkpoint inhibitor therapy. For example, melanoma of the skin which had previously been resistant to immunotherapy became sensitive when patients were given a fecal microbial transplant (FMT).

FMT may be refined by selection of strains based on their potential for immune stimulation. In one report, a consortium of bacterial strains that occur naturally in the gut of some but not all humans was found to enhance the immune response to cancer cells and infections.[17] In addition, certain strains of *Bifidobacterium* spp., such as *Bifidobacterium bifidum, Bifidobacterium pseudolongum,* and *Bifidobacterium longum* subsp. *longum,* modulate the immune response to cancer and are associated with improved outcomes to cancer immunotherapy. The effect is strain dependent and linked to strains with high levels of cell wall peptidoglycan.

In addition, translocation of the commonly used probiotic *Lactobacillus reuteri* to melanomas has been shown in mice, where it leads to upregulation of an antitumor $CD8^+$ T-cell response by metabolizing dietary tryptophan to indole-3-aldehyde which is a ligand of aryl hydrocarbon receptor (AHR), thereby facilitating immune checkpoint inhibitor therapy.[18]

While many of these strategies await conclusive evidence and regulatory approval, a more immediately available adjunct to immunotherapy is dietary intervention. Thus, recent studies have shown that high fiber diets with omega-3 consumption, such as with a Mediterranean diet, are associated with enhanced responses to immunotherapy. Moreover, higher immune response rates and less severe toxicities were correlated with a *Ruminococcaceae, Akkermansia muciniphilia* and methanogenic-archaea-dominated microbiome.[19]

Key points – cancer and the microbiome

- Chronic infections with specific microbes account for about 13% of global cancer cases and may accelerate progression of other cancers.
- The relationship between microbes and susceptibility to cancer is complex, variable and multifactorial.
- By shaping host immunity, an individual's personal microbiota modifies the progression of cancer.
- Intratumoral bacteria may have varying immunostimulatory or immunosuppressive effects on the host immune response.
- Immune stimulation by the gut microbiota is probably insufficient alone to eradicate most established cancers, but its contribution has become particularly evident in the setting of immunotherapy, particularly with checkpoint inhibitors.
- Modifying the microbiota by dietary means or by FMT has yielded encouraging results in combination with immune checkpoint inhibition.

References

1. Sepich-Poore GD, Zitvogel L, Straussman R, Hasty J, Wargo JA, Knight R. The microbiome and human cancer. *Science*. 2021;371(6536):eabc4552.
2. Scott AJ, Alexander JL, Merrifield CA, et al. International Cancer Microbiome Consortium consensus statement on the role of the human microbiome in carcinogenesis. *Gut*. 2019;68:1624–1632.
3. Cheng WY, Wu C-Y, Yu J. The role of gut microbiota in cancer treatment: friend or foe? *Gut*. 2020;69:1867–1876.
4. Saftien A, Puschhof J, Elinav E. Fungi and cancer. *Gut*. 2023;72:1410–1425.
5. Coleman OI, Haller D. Microbe–mucus interface in the pathogenesis of colorectal cancer. *Cancers*. 2021;13:616.

6. Baliou S, Adamaki M, Spandidos DA, Kyriakopoulos AM, Christodoulou I, Zoumpourlis V. The microbiome, its molecular mechanisms and its potential as a therapeutic strategy against colororectal carcinogenesis. *World Acad Sci J.* 2019;1:3–19.
7. Queen J, Shaikh F, Sears CL. Understanding the mechanisms and translational implications of the microbiome for cancer therapy innovation. *Nat Cancer.* 2023;4:1083–1094.
8. Puschhof J, Sears CL. Microbial metabolites damage DNA. *Science.* 2022;378:358–359.
9. Knippel RJ, Sears CL. The microbiome colorectal cancer puzzle: initiator, propagator, and avenue for treatment and research. *J Natl Compr Canc Netw.* 2021;19:986–992.
10. Blaser MJ. Disappearing microbiota: Helicobacter pylori protection against esophageal adenocarcinoma. *Cancer Prev Res (Phila).* 2008;1:308–311.
11. Lee KA, Thomas AM, Bolte LA, et al. Cross-cohort gut microbiome associations with immune checkpoint inhibitor response in advanced melanoma. *Nat Med.* 2022; 28: 535–544.
12. Choi Y, Lichterman JN, Coughlin LA, et al. Immune checkpoint blockade induces gut microbiota translocation that augments extraintestinal antitumor immunity. *Sci Immunol.* 2023;8: eabo2003.
13. Riquelme E, Zhang Y, Zhang L, et al. Tumor microbiome diversity and composition influence pancreatic cancer outcomes. *Cell.* 2019;178: 795–806.
14. Kalaora S, Nagler A, Wargo J, Samuels Y. Mechanisms of immune activation and regulation: lessons from melanoma. *Nat Rev Cancer.* 2022;22:195–207.
15. Kalaora S, Nagler A, Nejman D, et al. Identification of bacteria-derived HLA-bound peptides in melanoma. *Nature.* 2021;592:138–143.
16. Wallace BD, Wang H, Lane KT, et al. Alleviating cancer drug toxicity by inhibiting a bacterial enzyme. *Science.* 2010;330:831–835.
17. Tanoue T, Morita S, Plichta DR, et al. A defined commensal consortium elicits CD8 T cells and anti-cancer immunity. *Nature.* 2019;565:600–605.
18. Bender MJ, McPherson AC, Phelps CM, et al. Dietary tryptophan metabolite released by intratumoral Lactobacillus reuteri facilitates immune checkpoint inhibitor treatment. *Cell.* 2023;186: 1846–1862.
19. Simpson RC, Shanahan ER, Batten M, et al. Diet-driven microbial ecology underpins associations between cancer immunotherapy outcomes and the gut microbiome. *Nat Med.* 2022; 28: 2344–2352.

Gastroenterology

9 Drugs and the microbiota

HEALTHCARE

A common and clinically important interaction

The interaction between drugs and the microbiota is bi-directional; the microbiota modifies (activates or inactivates) many drugs, and many commonly prescribed drugs are associated with alterations of the composition of the microbiota (Box 9.1).[1-4] In addition to the direct actions of the microbiota on drug metabolism, indirect effects of the microbiota on the immune system alter the response to immunotherapy in patients with cancer (as discussed in Chapter 8) and chronic inflammatory disorders. The microbiota may also indirectly modify the metabolism of drugs by influencing the expression of hepatic enzymes such as the cytochrome P450 super family.

Drug–microbe associations have been detected in multiple human cohorts. In one study of 41 drugs, about 50% were linked in some way with the microbiome, most clearly in the case of proton pump inhibitors (PPIs), metformin, laxatives, as well as antibiotics.[5] Other microbiome-associated drugs included lipid-lowering statins,

---

BOX 9.1

Interactions between drugs and the microbiota

- Bi-directional
- Involve many commonly prescribed drugs
- A complex relationship confounded by co-medication and polypharmacy
- Includes direct effects on drug metabolism and indirect effects e.g. via the immune system
- May account in part for interindividual variations in drug responsiveness and toxicity
- May soon become part of the regulatory assessment for drug approval
- Drug-microbe interactions challenge efforts to define the normal human microbiome
- Confound studies of causality for human disease-associated microbiota
- Other xenobiotics, e.g. environmental or chemical pollutants and pesticides, may either modify or be metabolized by gut microbes

beta-blockers, ACE inhibitors, selective serotonin reuptake inhibitors, antidepressants, and other psychotropics. As previously discussed in Chapter 8, many cancer drugs inhibit microbial growth while simultaneously being degraded or modified by some components of the microbiota.

A study with over 1000 non-antibiotic drugs screened in vitro against 40 human gut microbial strains found that 1 in 4 drugs inhibited microbial growth.[6] In another study of 271 drugs incubated with gut microbes, 176 were metabolized to such a degree that the drug level dropped by over 20%.[7]

Mechanisms

For most drug-microbe associations, a mechanistic explanation is lacking. Untangling the interaction between drugs and the microbiome is complex. Firstly, the microbiome consists of 100–200-fold more genes than the human genome and correspondingly has a multitude of genes involved in regulating metabolism. Secondly, there is a high degree of interindividual variability of human gut microbes in the population. Thirdly, current understanding of gut microbiomes is largely limited to studies of humans living in North America and Europe but this accounts for a small minority of the global population and understanding of most ethnic groups is poor. Fourthly, there is no single common mechanism that underpins how drugs and microbes interact. While biotransformation (including activation and inactivation) of drugs by microbes has received much attention, drug accumulation by gut bacteria also appears to be a common mechanism that alters drug availability, responsiveness, pharmacokinetics and side effects. Thus, understanding drug-microbe interactions requires a case-by-case approach. Relatively few clinically important examples are understood[8-14] (Figure 9.1).

Progress in understanding the molecular basis and bi-directional nature of drug-microbe interactions is exemplified by the story of L-Dopa, which is administered orally to treat Parkinson's disease.[8] L-Dopa is absorbed in the gut, crosses the blood-brain barrier and is converted in the brain to dopamine by aromatic amino acid decarboxylase (AADC). However, variations in gut microbial activities may account for the highly variable clinical efficacy and toxicity of the drug (Figure 9.2).

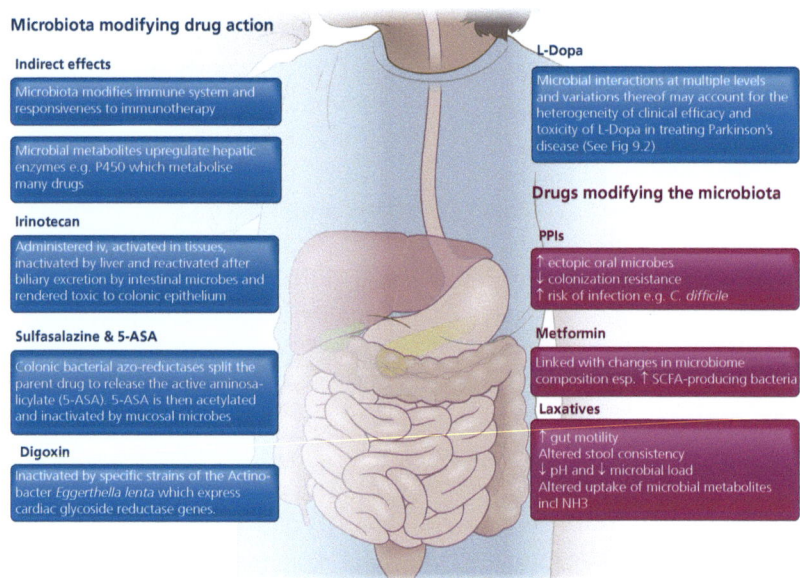

Figure 9.1 Bugs on drugs and drugs on bugs. Representative examples of gut microbial alteration of drug actions (blue) and examples of the influence of commonly used drugs on the gut microbiota (pink).

Figure 9.2 The gut microbiota in the treatment of Parkinson's disease with L-Dopa. L-Dopa, but not dopamine, is absorbed into the blood stream, crosses the blood-brain barrier and is converted to the therapeutically active agent dopamine by aromatic amino acid decarboxylase (AADC) in the brain. Since dopamine itself cannot cross the blood-brain barrier, L-Dopa is often co-prescribed with carbidopa, an inhibitor of peripheral AADC, or with entacapone, a reversible catechol-*O*-methyltransferase (COMT) inhibitor that prevents conversion to dopamine outside the brain. However, the microbiota plays a role in modulating the drug's actions at 3 levels: the microbial decarboxylase tyrosine decarboxylase (TyrDC) which is expressed by *E. faecalis* can also metabolize L-Dopa and reduce its bioavailability; further metabolism by dopamine-dehydroxylase found in *Eggerthella lenta* and other bacteria can convert dopamine to m-tyramine which is associated with side effects including hypertensive crises; and additional bi-directional interactions occur between carbidopa and entacapone which modify the microbiota and are also metabolized by the microbiota.

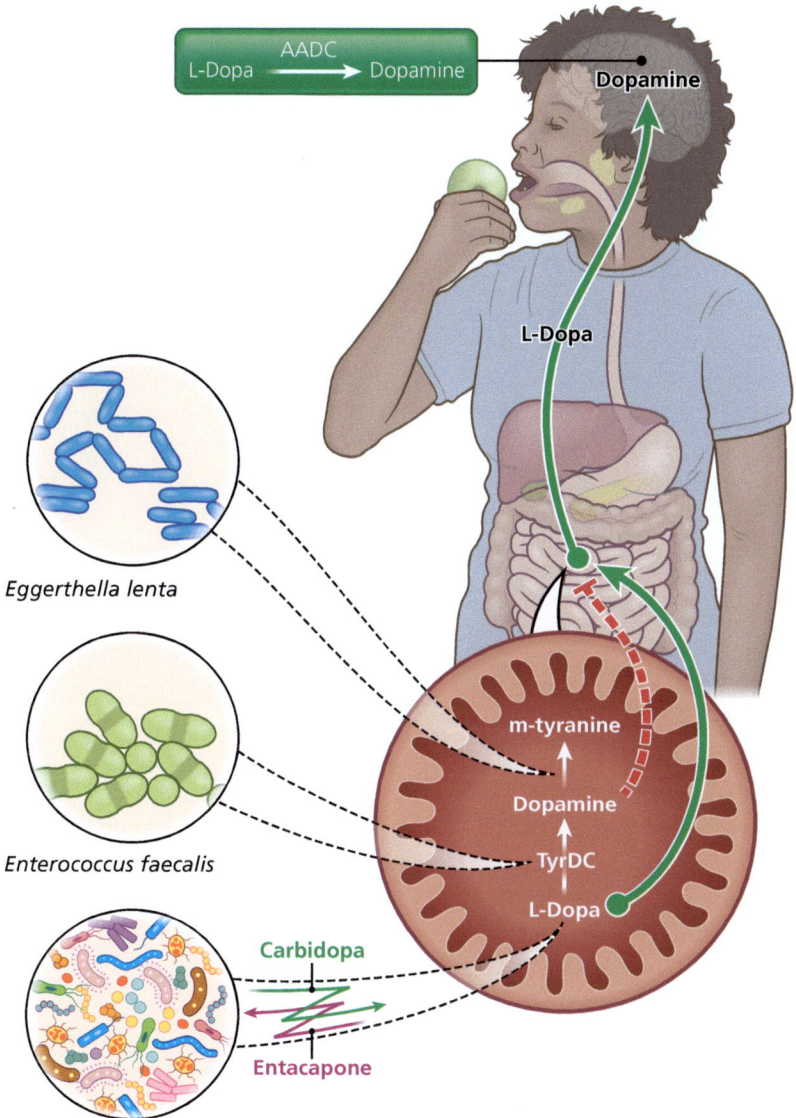

In many cases, microbe-drug interactions are both direct and indirect. For example, while PPIs indirectly affect the microbiota by suppressing gastric acid, they also have direct effects by inhibiting certain bacteria such as *Dorea* and *Ruminococcus* species.[1,15,16]

Laxatives of all types affect the composition and microbial load in the gut by accelerating transit and altering stool volume and consistency. The dual action of lactulose, a galactose-fructose disaccharide, is used in the management of hepatic encephalopathy. It is metabolized in the colon by saccharolytic bacteria to lactic acid and small amounts of formic and acetic acids which increase gut motility and exert an osmotic effect to increase water content of stool, thereby increasing the frequency of bowel movements. Simultaneously, by acidifying the lumen it enables ammonia (NH_3) to diffuse into the colon where it becomes ionized to ammonium ions (NH_4^+) that cannot be absorbed into the blood. Acidification of the colonic contents also constrains urease-producing bacteria and ammonia formation.

Curiously, transient and mild osmotic diarrhea which may be secondary to malabsorption or induced by laxatives, such as polyethylene glycol (PEG), has been associated with microbiome disturbances which in experimental mice are long-lasting.[2,17]

Alcohol and tobacco

Several studies have implicated the gut microbiota and intestinal bacterial overgrowth in the manifestations of alcohol-related liver disease and there have been numerous reports of alterations in the microbiome in chronic alcoholism. However, alcohol is mainly absorbed in the stomach and upper small bowel with limited access to the microbiota in the distal bowel. Experiments in mice suggest that ethanol is not metabolized by the gut microbiota but that alterations in the microbiota are secondary to alcohol-induced acetate production by the liver which serves as a source of increased carbon for bacterial growth in the intestine.[18]

As with alcohol, it is difficult to find any physiological process that is unaffected by cigarette smoking. Cigarette smoking and cessation of smoking have been associated with changes in the composition of the microbiota particularly in rodents, although studies in humans have been inconsistent. Recently, a mechanistic link between the cessation of smoking and microbiome-induced weight gain has been shown in a murine model.[19] Unexpectedly, it appears that non-nicotine components of tobacco cause the remodeling of the gut microbiota to enhance energy harvest from food and promote weight gain. In contrast, nicotine is metabolized by the gut microbiota; in particular, *Bacteroides xylanisolvens* has been identified as a nicotine degrader.[20]

The microbiota as a repository for drug discovery

While the role of the microbiome may soon become a mandatory consideration for drug development and approval by regulatory agencies, the microbiota also represents a repository of bioactive metabolites. Mankind already uses microbial toxins such as botulinum toxin (Botox) for clinical and cosmetic purposes. However, the microbial metabolites mediating host-microbe and microbe-microbe interactions have also been mined using culture-based, metagenomics-based and metabolomics-based strategies. Bioactives with potential therapeutic utility include bacteriocins with narrow-spectrum antimicrobial activity, bioactive amino acid metabolites, immunoregulatory polysaccharides and glycolipids, anti-inflammatory peptidoglycans and growth factors.[21,22] Proof of principle for the clinical application of several microbial-derived bioactives has been achieved. For example, the discovery of a bifidobacterial pilus protein (TadE) with growth factor-like activity for intestinal epithelial proliferation may underpin the protective effects of a bifidobacterial probiotic against epithelial ulceration caused by non-steroidal anti-inflammatory drugs.[23]

Key points – drugs and the microbiota

- Microbe-drug interactions are bi-directional; the microbiota either directly or indirectly modifies (activates or inactivates) many drugs, and many commonly prescribed drugs have been associated with alterations to the microbiota.
- Interindividual variations in drug responsiveness and toxicity may be due in part to microbe-drug interactions.
- Common examples of drugs that modify the microbiota include metformin, PPIs, and laxatives.
- Common examples of drugs whose actions are modified by the microbiota include irinotecan, digoxin, L-Dopa, sulfasalazine and aminosalicylates.
- Alcohol and tobacco have complex and indirect effects on the microbiota.
- The microbiota also represents a repository of bioactive metabolites, some of which may be mined for novel drug discovery.

References

1. Jackson MA, Verdi S, Maxan ME, et al. Gut microbiota associations with common diseases and prescription medications in a population-based cohort. *Nat Commun.* 2018; 9: 2655.
2. Weersma RK, Zhernakova A, Fu J. Interaction between drugs and the gut microbiome. *Gut* 2020;69:1510–1519.
3. Subramaniam S, Kamath S, Ariaee A, Prestidge C, Joyce P. The impact of common pharmaceutical excipients on the gut microbiota. *Expert Opin Drug Deliv.* 2023;20:1297–1314.
4. Guthrie L, Kelly L. Bringing microbiome-drug interaction research into the clinic. *EBioMedicine* 2019;44:708–715.

5. Vich Vila A, Collij V, Sanna S, et al. Impact of commonly used drugs on the composition and metabolic function of the gut microbiota. *Nat Commun.* 2020;11:362.
6. Maier L, Pruteanu M, Kuhn M, et al. Extensive impact of non-antibiotic drugs on human gut bacteria. *Nature.* 2018;555: 623–628.
7. Zimmermann M, Zimmermann-Kogadeeva M, Wegmann R, et al. Mapping human microbiome-drug metabolism by gut bacteria and their genes. *Nature.* 2019; 570:462–467.
8. Rekdal VM, Bess EN, Bisanz JE, et al. Discovery and inhibition of an interspecies gut bacterial pathway for levodopa metabolism. *Science.* 2019;364: eaau6323.
9. Forslund K, Hildebrand F, Nielsen T, et al. Disentangling type 2 diabetes and metformin treatment signatures in the human gut microbiota. *Nature.* 2015; 528: 262–266.
10. Silpe JE, Balskus EP. Deciphering human microbiota – host chemical interactions. *ACS Cent Sci.* 2021;7:20–29.
11. Woo, AYM, Aguilar Ramos MA, Narayan R, et al. Targeting the human gut microbiome with small molecule inhibitors. *Nat Rev Chem.* 2023;7:319–339.
12. Lindell AE, Zimmermann-Kogadeeva M, Patil KR. Multimodal interactions of drugs, natural compounds and pollutants with the gut microbiota. *Nat Rev Microbiol.* 2022;20:431–443.
13. Mehta RS, Mayers JR, Zhang Y, et al. Gut microbial metabolism of 5-ASA diminishes its clinical efficacy in inflammatory bowel disease. *Nat Med.* 2023; 29: 700–709.
14. Klünemann M, Andrejev S, Blashe S, et al. Bioaccumulation of therapeutic drugs by human gut bacteria. *Nature.* 2021;597(7877):533–538.
15. Wan Y, Zuo T. Interplays between drugs and the gut microbiome. *Gastroenterol Rep (Oxf).* 2022;10:goac009.
16. Zhang J, Zhang C, Zhang Q, et al. Meta-analysis of the effects of proton pump inhibitors on the human gut microbiota. *BMC Microbiol.* 2023;23:171.
17. Tropini C, Moss EL, Merrill BD, et al. Transient osmotic perturbation causes long-term alteration to the gut microbiota. *Cell.* 2018;173:1742–1754.
18. Martino C, Zaramela, L.S, Gao B, et al. Acetate reprograms gut microbiota during alcohol consumption. *Nat Commun.* 2022;13:4630.
19. Fluhr L, Mor U, Kolodziejczyk AA, et al. Gut microbiota modulates weight gain in mice after discontinued smoke exposure. *Nature.* 2021; 600:713–719.
20. Chen B, Sun L, Zeng G, et al. Gut bacteria alleviate smoking-related NASH by degrading gut nicotine. *Nature.* 2022;610: 562-568.

21. Wang L, Ravichandran V, Yin Y, Yin J, Zhang Y. Natural products from mammalian gut microbiota. *Trends Biotech*. 2019;37:492–504.
22. Donia MS, Fischbach MA. Small molecules from the human microbiota. *Science*. 2015;349:395.
23. Mortensen B, Murphy C, O'Grady J, et al. Bifidobacterium breve Bif195 protects against small intestinal damage caused by acetylsalicylic acid in healthy volunteers. *Gastroenterology*. 2019;157:637–646.

10 Antimicrobial resistance

HEALTHCARE

Definition and development

Antimicrobial resistance (AMR) occurs when microorganisms including bacteria, viruses, parasites and fungi become resistant to antimicrobial treatments to which they were previously susceptible. To understand AMR there are some old truths that need to be acknowledged (Box 10.1). Antimicrobials and AMR predated humans on the planet.[1-4] While a low level of AMR in the environment may be anticipated as an evolutionary response by microbes, human activities have escalated the spread of AMR.[5,6] Increased rates of AMR have been compounded by the diminishing pipeline of development of new antimicrobial drugs (Figure 10.1).

Although misuse of antimicrobials in human healthcare receives much attention and even increased during the Covid-19 pandemic,[7] antibiotic use in farming is the major contributor to AMR and is projected to grow despite efforts to curtail it.[5] Antibiotics are often used to enhance animal growth and prevent diseases in animals kept in crowded, unsanitary conditions. Much of this is not officially recorded. China uses more antibiotics in farming than any other country, with African countries and Pakistan projected to rapidly increase their consumption.[5]

BOX 10.1

Old truths regarding antimicrobial resistance (AMR)

- Microbes not humans 'invented' antibiotics and AMR
- Resistance predated humans on the planet
- Resistance exists to drugs not yet developed
- Appropriate as well as inappropriate use of antibiotics selects for resistance
- Human healthcare accounts for only a minority of AMR
- Eliminating inappropriate use will not solve the problem
- Resistance is transmissible: it behaves like an infectious disease
- Public and political understanding of AMR is poor and inaccurate

Antimicrobial resistance

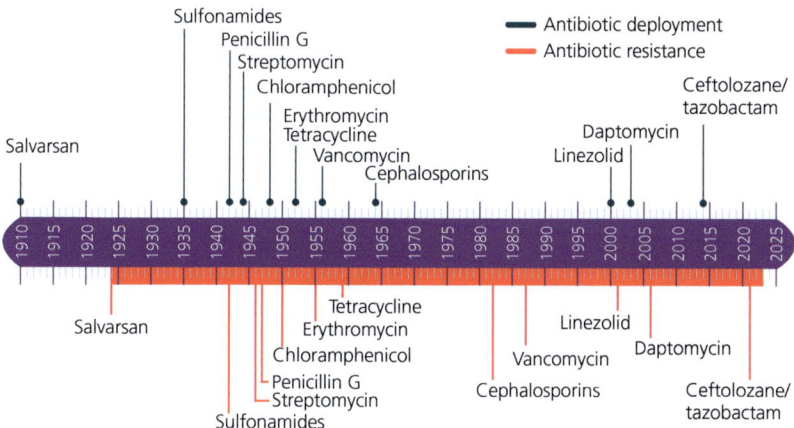

Figure 10.1 Relentless emergence of AMR following commercial deployment of antimicrobial agents. Adapted from Figure 1, United Nations Environment Programme 2023.[6]

In addition to consumption by humans and in farming, environmental drivers also play a significant role in the development, transmission and spread of AMR, including back to humans and animals.[6,8] Co-prescription of non-antibiotic drugs may also contribute to the development of AMR in humans.[9]

Health, social and economic consequences

Anti-microbial resistance is a global problem and is spreading like a pandemic. It is a serious threat to human, animal and planetary health with additional social and economic implications and one of the top 10 global threats to human health according to the WHO. Estimates suggest that AMR may be responsible for up to 10 million deaths globally by 2050 – a death toll similar to that caused by cancer. Low-income countries and those with a major burden of infectious disease in Africa and Asia will be disproportionately affected. Because AMR affects the health of animals and plants, economic development will be retarded, and social inequities amplified. Unless new antimicrobial strategies are developed, routine healthcare procedures including surgery, cancer therapy, immunotherapy, and others will become jeopardized in all countries (Box 10.2).[11–13]

> **BOX 10.2**
>
> **View of a post-antibiotic world**
>
> - Routine surgery becomes high risk
> - Increased hospitalizations for infections
> - Routine obstetric care becomes high risk
> - Healthcare costs rise
> - Reversed trends for life expectancy globally
> - Annual gross domestic product falls significantly
> - Low-income countries worst affected
> - Poverty rates rise globally

Sources: WHO and Jonas et al. 2017.[10]

The WHO has prioritized a list of pathogens for which there is particular need for new research and development of new antibiotics (Table 10.1).

Environmental dimension

While attention has understandably and correctly been directed toward the human gut as a reservoir of AMR, the main reservoir of antibiotic-resistant microbes is not human but is the environment – soil, water and animals – that harbors most of the world's resistant microbes, contaminated because of human activity. The intrinsic and acquired AMR genes collectively carried by microorganisms in an environmental niche are referred to as that environment's *resistome*.[4] Release of antimicrobials into the environment selects for resistant microorganisms and increases the likelihood of the emergence of resistance. Spread of AMR is greatly increased by flooding of contaminated sources, agricultural runoff, airborne transmission, wildlife migration, globalization, climate change and by human and wildlife migration. The presence of heavy metals and other chemicals increases the risk of resistance emerging.

TABLE 10.1
WHO priority list of pathogens targeted for new antibiotic development

Priority 1: Critical

Acinetobacter baumannii, carbapenem-resistant

Pseudomonas aeruginosa, carbapenem-resistant

Enterobacteriaceae, carbapenem-resistant, ESBL-producing

Priority 2: High

Enterococcus faecium, vancomycin-resistant

Staphylococcus aureus, meticillin (methicillin)-resistant, vancomycin-intermediate and resistant

Helicobacter pylori, clarithromycin-resistant

Campylobacter spp., fluoroquinolone-resistant

Salmonellae, fluoroquinolone-resistant

Neisseria gonorrhoeae, cephalosporin-resistant, fluoroquinolone-resistant

Priority 3: Medium

Streptococcus pneumoniae, penicillin-non-susceptible

Haemophilus influenzae, ampicillin-resistant

Shigella spp., fluoroquinolone-resistant

ESBL: extended spectrum beta lactamase.
Source: who.int/news/item/27-02-2017-who-publishes-list-of-bacteria-for-which-new-antibiotics-are-urgently-needed, last accessed 5 March 2024.

Human and animal exposure to environmental AMR may occur following consumption of contaminated food or water.[4,6]

Mechanisms and transmission of resistance

Antibiotic resistance mechanisms operate at various levels, including:
- restricting uptake of antibiotics (this is of particular relevance to Gram-negative organisms)
- modifying of the antibiotic target
- antibiotic inactivation
- active antibiotic efflux pump.

AMR may be intrinsic or acquired by mutations or (in the case of bacteria) by the acquisition of resistance genes from different microbes by horizontal gene transfer (HGT) of mobile genetic elements.[14,15] HGT is common among gut bacteria. Some bacteria may take up free DNA from the environment and incorporate it into their chromosome (transformation). Genes may also be transferred horizontally as plasmids and transposons by direct contact with recipient cells (conjugation). In some cases, a bacteriophage may transfer fragments of bacterial DNA including the resistance sequence between bacteria (transduction). In addition, a role for membrane vesicles in the horizontal transmission of DNA among gut bacteria has been identified (Figure 10.2).

Communicating the problem

The understanding of AMR by the public and among healthcare policy makers is often poor and inaccurate.[16] Surveys in Europe and the USA have shown that many people think that AMR means that an antimicrobial drug has stopped working for them rather than reflecting a property of the microbe for which the antimicrobial agent was prescribed. Equally concerning is the failure of policy makers and others to appreciate that AMR is contagious, an issue which has immediate relevance to hospital design and overcrowding. Similarly, public health messaging regarding AMR has been ineffectual, abstract, impersonal and lacking immediacy. Although exhortations to constrain injudicious use of antibiotics are appropriate, they could be more effective if supplemented with an explanation of the adverse effects of antimicrobial agents on the normal microbiome.

Figure 10.2 Mechanisms of resistance and horizontal gene transfer of resistance. Transformation: some bacteria can take up naked DNA from the environment. Membrane vesicle fusion: lipid bilayer-enclosed vesicles can transport DNA and other material between bacteria. Transduction: genetic material can be transferred between donor and recipient bacteria via a bacteriophage. Conjugation: mobile genetic elements, such as plasmids, can transfer via a pilus formed between donor and recipient cells. Adapted from McInnes et al. 2020.[14]

Antimicrobial resistance

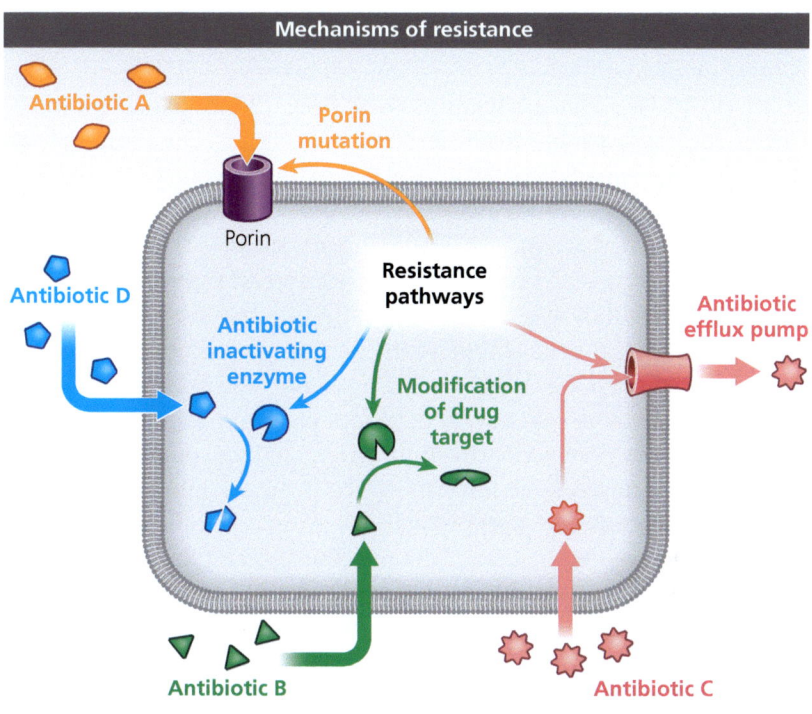

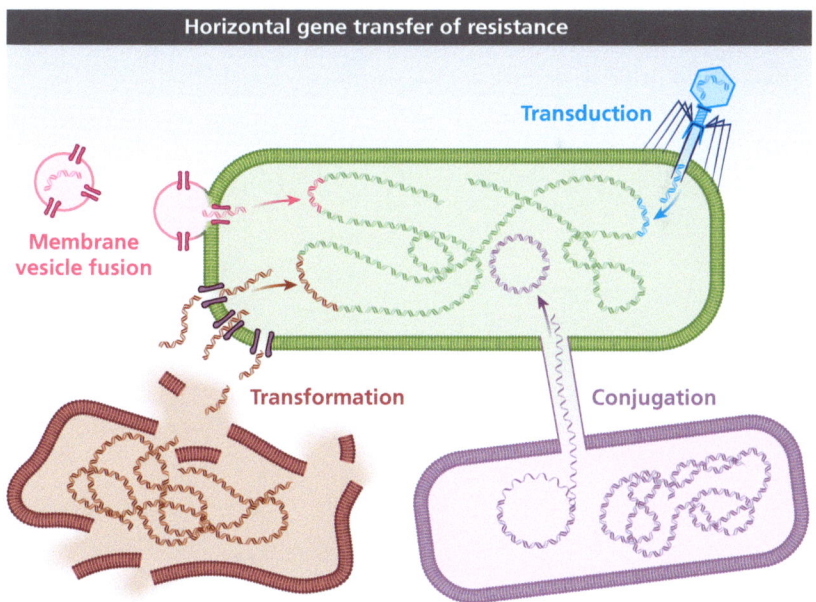

A 'One Health' approach to a 'One Health' problem

Preventing the spread of AMR is a multidimensional global problem. 'One Health' is a concerted approach at global, regional and country levels with engagement from all sectors, stakeholders and institutions. Simply stated, this approach acknowledges interdependence of the health of people, animals, plants and the environment and requires integrated unified attention by government, society, international organizations and the private sector. The traditional response to AMR is a triad of antibiotic stewardship to constrain misuse, infection prevention to eliminate the need for antimicrobials, and new antimicrobial development.[2,17,18] Each of these must continue. However, infections will always occur despite rigorous preventive strategies. Moreover, while the diminishing pipeline of antimicrobial development might be restored by correcting economic and regulatory barriers and perhaps by use of artificial intelligence, AMR is inevitable even with appropriate use of antimicrobials. It has even been shown that antibiotic resistance to drugs not yet invented is already widespread in nature (see Box 10.1, page 108). Therefore, innovative approaches, such as mobilizing the protective properties of the microbiome, enhancing immune responses against infectious agents, use of monoclonal antibodies against microbes, use of alternative agents including phages, and designing new antimicrobials that do not drive resistance are called for.

Key points – antimicrobial resistance

- AMR is a global problem and a major threat to human health, economic stability and planetary health.
- Antimicrobials and AMR are naturally occurring and predated humans on the planet, but human activities have accelerated the spread of AMR.
- Increased AMR is compounded by a diminishing pipeline of new antimicrobial agents.
- Resistance to antimicrobials is transmissible, spreading like an infectious disease.
- Public and political understanding of AMR is poor and often inaccurate.
- Tackling AMR requires a 'One Health' approach – concerted global and regional engagement from all sectors of all societies, including innovative approaches such as mobilizing the microbiome against infectious threats.

References

1. Larsen J, Raisen CL, Ba X, et al. Emergence of methicillin resistance predates the clinical use of antibiotics. *Nature.* 2022;602(7895):135–141.
2. Spellberg B, Bartlett JG, Gilbert DN. The future of antibiotics and resistance. *N Engl J Med.* 2013;368:299–302.
3. Landecker H. Antibiotic resistance and the biology of history. *Body Soc.* 2016;22:19–52.
4. Perry J, Waglechner N, Wright G. The prehistory of antibiotic resistance. *Cold Spring Harb Perspect Med.* 2016;6:a025197.
5. Reardon S. Antibiotic use in farming set to soar despite drug resistance fears. *Nature.* 2023;614:397.

6. United Nations Environment Programme. Bracing for superbugs: strengthening environmental action in the One Health response to antimicrobial resistance. Geneva: UNEP, 2023. unep.org/resources/superbugs/environmental-action, last accessed March 5 2024.
7. Al Sulayyim HJ, Ismail R, Hamid AA, Ghafar NA. Antibiotic resistance during COVID-19: A systematic review. *Int J Environ Res Public Health*. 2022;19:11931.
8. Zhou Z, Shuai X, Lin Z, et al. Association between particulate matter (PM)2–5 air pollution and clinical antibiotic resistance: a global analysis. *Lancet Planet Health*. 2023;7:e649–e659.
9. Wang Y, Yu Z, Ding P, et al. Antidepressants can induce mutation and enhance persistence toward multiple antibiotics. *Proc Natl Acad Sci USA*. 2023;120(5):e2208344120.
10. Jonas OB, Irwan AB, Le Gall FCJ, Marquez FG, Patricia V. Drug-resistant infections: a threat to our economic future (Vol. 2) : final report (English). Washington D.C.: HNP/Agriculture Global Antimicrobial Resistance Initiative; World Bank Group, 2017. documents.worldbank.org/curated/en/323311493396993758/final-report.
11. Fair RJ, Tor Y. Antibiotics and bacterial resistance in the 21st century. *Perspect Medicin Chem*. 2014;6:25–64.
12. Iskandar K, Murugaiyan J, Hammoudi Halat D et al. Antibiotic discovery and resistance: the chase and the race. *Antibiotics*. 2022;11:182.
13. Murray CJI, Ikuta KS, Sharara F, and the Antimicrobial Resistance Collaborators. Global burden of bacterial antimicrobial resistance in 2019: a systematic analysis. *Lancet*. 2022;399(10325): 629–655.
14. McInnes RS, McCallum GE, Lamberte LE, van Schaik W. Horizontal transfer of antibiotic resistance genes in the human gut microbiome. *Curr Opini Microbiol*. 2020; 53:35–43.
15. Anthony WE, Burnham CD, Dantas G, Kwon JH. The gut microbiome as a reservoir for antimicrobial resistance. *J Infect Dis*. 2021;223 (12 Suppl 2):S209-S213.
16. Mendelson M, Balasegaram M, Jinks T, Pulcini C, Sharland M. Antibiotic resistance has a language problem. *Nature*. 2017; 545: 23–25.
17. Rawson TM, Wilson RC, O'Hare D, et al. Optimizing antimicrobial use: challenges, advances and opportunities. *Nat Rev Microbiol*. 2021;19(12): 747–758.
18. Mitcheltree MJ, Pisipati A, Syroegin EA, et al. A synthetic antibiotic class overcoming bacterial multidrug resistance. *Nature*. 2021;599(7885): 507–512.

Gastroenterology

11 The gut virome and mycobiome

HEALTHCARE

The gut virome

Long considered part of the 'dark matter' in the gut microbiome because of technological constraints on its exploration, the gut virome is still under-explored. However, knowledge of gut viruses has improved in recent years (Box 11.1).[1-4] Most gut viruses are thought to be phages.[5-7] The major taxa of phages are typically *Caudovirales* (tailed phages) and *Microviridae* (icosahedral non-tailed phages). However, as new sequence-based data emerge, the taxonomy of the gut virome is likely to be revised.[8]

Most gut viruses characteristically lack an envelope (except coronaviruses) probably because a lipid envelope would be susceptible to the detergent effect of bile salts or to dehydration in the distal colon.

BOX 11.1

Properties of the human gut virome

- Site-specific variability in composition across anatomic niches (e.g. gut, skin, respiratory tract)
- Assembled early during first 2 years and stabilizes in healthy adulthood
- High interindividual variability related to environment, ethnicity, diet, and age
- Gut is repository of greatest number of viruses in the human body ($\sim 10^9$–10^{10} virus-like particles/g feces) similar to that of gut bacteriome
- Composed of single-stranded and double-stranded DNA and RNA viruses
- Includes mainly viruses that infect bacteria and archaea (phage; >90% of total) and viruses that infect host cells (<10% of total; including predominantly latent herpes viruses, anelloviruses and adenoviruses) and transient viruses in food
- Phageome diversity varies inversely with bacteriome diversity (shifting from high phageome/low bacteriome diversity at birth to the opposite after 2 years)
- Trans-kingdom interconnectivity with host immunity and the bacteriome

The functional impact of the virome on mucosal homeostasis relates to its reciprocal interactions with the bacteriome and with the mucosal immune system.[7] By shaping the composition of the microbiota, phages indirectly influence host immunity and metabolism, whereas the functionality and makeup of the virome is responsive to the bacteriome and the host immune system (Figures 11.1 and 11.2).

Phage-bacteriome interaction

Phages have variable interactions with bacteria depending on their replication cycle. *Lytic* phage replicate by injecting their nucleic acid into the bacterium, leading to synthesis of new viruses, lysis and release of multiple replicas (progeny phage). In contrast, with *temperate* phage (also known as *lysogenic* phage) the injected DNA becomes integrated into the bacterial chromosome forming a *prophage*, which remains quiescent until it is excised and switches to lytic growth if exposed to environmental stressors (Figure 11.2). Phage also have other forms of interaction with bacteria: the phage genome

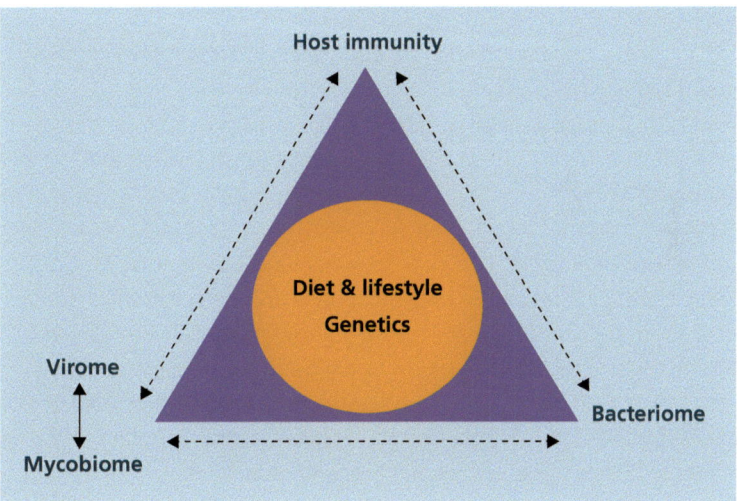

Figure 11.1 Trans-kingdom tripartite interconnectivity regulates homeostasis in the gut, each component of which is shaped by dietary, lifestyle and genetic variables.

may not integrate with the bacterial genome or replicate or cause lysis (*pseudolysogeny*) and some phages (e.g. filamentous *Plasmaviridae*) may infect the bacterium and produce new progeny by *budding* without causing lysis.[5,6]

In addition, phages may change the functional capacity and pathogenicity of commensal microbes. This may include facilitating biofilm formation, metabolism and antibiotic resistance by horizontal gene transfer.

Phage-immune interaction

Phages interact with the host immune system both directly and indirectly via the bacteriome and influence the inflammatory tone of the mucosal immune system.[5,6] Billions of phage particles are transported across the intestinal epithelium daily where they encounter dendritic cells and other cellular components of the mucosal immune response, thereby activating immune effector T and B cells and a cytokine cascade via Toll-like receptors (TLRs). This has been demonstrated by feeding an *E. coli* phage isolated from

Figure 11.2 Signaling among viral, bacterial and host immune cells contributes to interindividual variability in composition and functionality of the microbiome. While phage may theoretically constrain the numbers of fast-growing bacteria ('kill the winner'), evidence for this in the gut is limited. Alternatively, commensal bacteria may deploy the phage they carry to help compete with pathogenic competitors, whereas in health it seems likely that the predator-prey relationship between phage and bacteria remains stable, each keeping pace with one another to stand still ('Red Queen' dynamics). In addition, horizontal gene transfer by phage to the bacteriome may alter the functional capacity of bacteria (phenotypic change). Transition from a lysogenic to lytic state is influenced by various stressors including antibiotics and disease. While some viruses with immunoglobulin-like variable domains bind to mucin glycoproteins and enhance barrier protection, other viral proteins are transported across the mucosa and engage continually with Toll-like receptors on dendritic and other immune cells thereby shaping the mucosal immune and inflammatory tone.

(a) Virus-bacteriome interaction

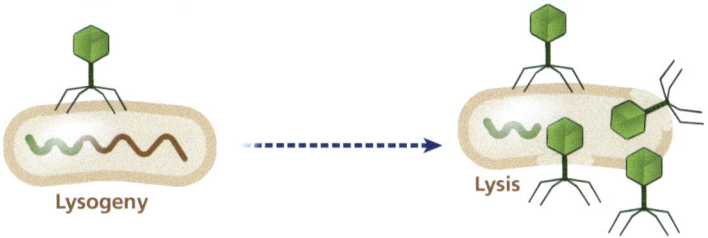

Predator-prey Dynamics
'Kill the winner'
Kill competitor
Red Queen Dynamics

Stressors
Antibiotics, pH, Oxidative Stress, Inflammation, and Disease

(b) Virus-immune engagement

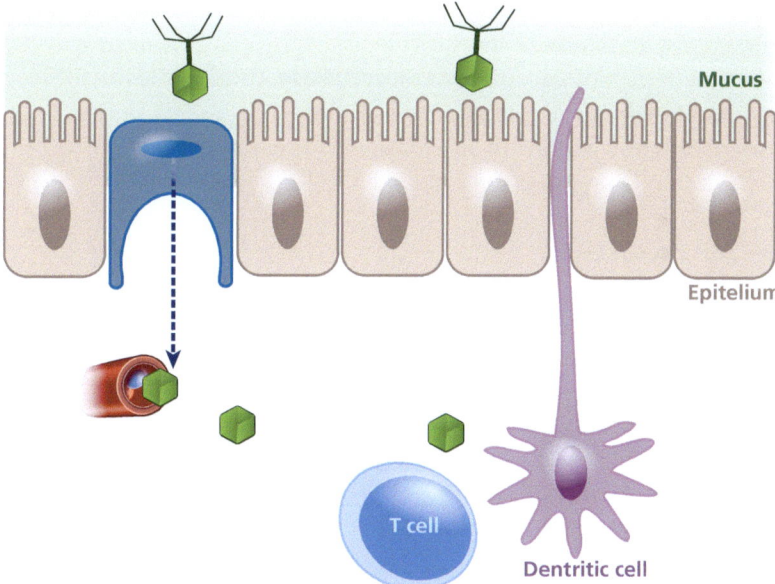

the human gut to germ-free mice which then increased interferon gamma (IFNγ)-producing mucosal T cells. When phages enter the circulation they are removed by phagocytes in the liver and spleen and neutralized by phage-specific antibodies. Phages produced by pathogenic bacteria may also be sensed by TLRs on immune cells and trigger a type 1 interferon response.

In addition, some phages contribute to host defense by expressing immunoglobin-like domains on their protein shell (capsid) which binds to mucin glycoproteins within intestinal mucus and may contribute to barrier protection against pathogens[9] (Figure 11.2b).

The virome and disease

Disturbances in phage counts and diversity have been associated with several chronic inflammatory and metabolic disorders. Whether this reflects a cause or consequence of disease is unclear because fluctuations in phages are expected in disorders where the bacteriome is disturbed. Moreover, viruses (including phage) may have both protective and adverse effects on the host.[6,7] Thus, in animal models, expansion of phage particles has been associated with exacerbation of colitis,[10] but separate evidence suggests that phages and eukaryotic viruses may compensate for the absence of bacteria and restore intestinal morphology and host immunity in gnotobiotic mice.[11,12] Little is known of the role of phages in human diseases but there is increasing interest in the use of phage for therapeutic purposes (discussed in Chapter 12).

The gut mycobiome

Fungi account for a tiny (~0.1%) but important component of the resident gut microbiota with a disproportionate influence on the host immune response and on the gut bacteriome (Box 11.2).[13] As the most direct route of entry for food and airborne spores, the oral cavity is colonized with a diversity of fungi, most notably *Candida,* but fungi are also detectable throughout the gastrointestinal tract in most adults.

In contrast to the bacteriome, methods of extraction and identification of fungi have not been rigorously standardized and different methods of processing may result in different relative abundances of species such as *Penicillium, Malassezia* and

Debaryomyces. In further contrast to most bacteria, fungi have a sturdy cell wall that may confound extraction procedures.

Like the virome and the bacteriome, the mycobiome varies with the age of the host and is influenced by dietary and lifestyle variables, particularly diet and exposure to antibiotics. Colonization with fungi begins at birth. As expected, vaginal delivery is associated with colonization by yeasts from the maternal vagina. Colonization by *Candida* spp. is, therefore, transmitted vertically but also horizontally from the environment.

Saccharomycetales and *Malasseziales* spp. dominate the postnatal gut mycobiome during the first few weeks of life, followed by a decline of *Malasseziales* with an abrupt change at weaning from

---

BOX 11.2

Features of the gut mycobiome

- Fungal genes account for ~0.1% of the resident gut microbiome
- Identified with sequencing techniques targeting the fungal ribosomal RNA gene cluster – 18S small subunit rDNA, 28S large subunit rDNA and the internal transcribed spacer (ITS1 or ITS2)
- Most commonly detected fungi in the human gut include *Candida albicans, C. tropicalis, C. parapsilosis, C. glabrata, C. krusei, S. cerevisiae, Cladosporium cladosporioides, Penicillium allii, Malassezia globosa, M. restricta, Debaryomyces hansenii* and *Galactomyces geotrichum*
- Many detected fungi are non-viable in the gut or in transit from dietary or environmental exposure without colonization
- Carbohydrate-rich diets are associated with *Candida* spp. in the gut, whereas vegetarian diets have greater abundance of spore-forming and dietary fungi (*Fusarium* spp. and *Penicillium* spp.). Fermented foods are a rich source of yeasts and filamentous fungi
- Broad-spectrum antibiotics disrupt bacterial-fungal interactions and are a risk factor for candidemia in seriously ill patients
- The host immune system expresses pattern recognition receptors, such as the C-type lectin Dectin-1, to recognize beta-glucans in fungal cell walls and deploys interleukins (IL) such as IL-17 and IL-22 to protect against invasive fungal infections

breast milk to solid food when *Saccharomyces cerevisiae* becomes the most abundant species along with *Cystofilobasidium* spp., *Ascomycota* spp., and *Monographella* spp., consistent with a prominent role for diet in shaping the gut mycobiome.[13] The widespread presence of *S. cerevisiae* in food and beverages may account for its frequent detection in the gut.

In adulthood, the diversity of the gut mycobiome increases with *Candida*, *Saccharomyces*, and *Cladosporium* being the most abundant genera. An increase of *Candida* spp. has been inversely correlated with the diversity of the bacterial microbiome and also associated with diseases of a Western lifestyle including obesity and inflammatory bowel disease. With advancing age (>65 years), *Penicillium*, *Candida*, *Saccharomyces*, and *Aspergillus* are the most common genera in the gut mycobiome.

Fungal interactions with the bacteriome

The discovery of penicillin highlighted the importance of fungal-bacterial interactions. In the gut, interactions with other microbes determine overall community function and include competition or collaboration for access to nutrients, changes in local environmental pH, production of metabolites inhibitory to competitors, formation of mixed biofilms and modification of virulence.

Several examples illustrate the importance of cross-kingdom interactions. Thus, the virulence of *C. albicans* can be enhanced by interacting with enterohemorrhagic *E. coli* or reduced by interacting with *Clostridioides difficile* and *Enterococcus faecalis*.[13,14] There is also an intriguing interaction between *C. albicans* and *H. pylori* in the stomach, whereby the bacterium may survive in vacuoles in the fungus, protected from gastric acid.

Fungal interactions with mucosal immunity

Fungal cell wall components are important drivers of mucosal cellular and humoral immunity, including Th17 immune responses, which restrict the overgrowth of gut fungi and the production of mucosal IgA antibodies, which suppress the hyphal form of fungi and its virulence factors.[14,15] Since *C. albicans* colonization is a potent inducer of secretory IgA, it indirectly regulates its own hyphal morphogenesis to maintain commensalism. Importantly, commensal fungi can replace the protective effects of the bacteriome. Thus, when commensal bacteria have been depleted by antibiotics in mice, their susceptibility to infection and inflammation may be offset by monocolonization with either *C. albicans* or *S. cerevisiae*. This is mediated by mannans and other fungal cell wall components which engage with C-type lectins, TLRs and other pattern recognition receptors on immune cells.

The mycobiome and disease

Candida species in the gut are considered to be a major cause of infections and invasive candidiasis. *C. albicans* is normally a harmless commensal but has the potential to become invasive if the gut microbiome is disrupted, if barrier function is disrupted, or if the host is immune suppressed. *Candida* overgrowth has been associated with diabetes, Crohn's disease and immune suppression. Recurrent mucocutaneous candidiasis is also associated with deficiency of Dectin-1 in humans.

Several lines of evidence point toward disturbances of the gut mycobiome in inflammatory bowel disease (IBD).[13]

- Genetic polymorphisms of Dectin-1 have been linked with increased severity of ulcerative colitis.
- Genetic polymorphisms of CARD9^{S12N} in patients with IBD are associated with the presence of *Malassezia* spp. in the gut mucosa. *M. restricta* was found to aggravate gut inflammation in mice and in humans with Crohn's disease who are homozygous for the CARD9^{S12N} polymorphism.
- *D. hansenii*, which is used in the food industry, was found to be increased in the lesions of Crohn's disease and is associated with impaired colonic healing.
- Serum anti-*S.cerevisiae* antibodies are a serological marker for a subset of Crohn's disease patients.
- *Candida* spp. drive Th17 and interleukin (IL)-23 responses and aggravate murine models of colitis.
- Curiously, antifungal treatment has been reported to exacerbate experimentally induced colitis in mice whereas monocolonization with *C. albicans* or *S. cerevisiae* is protective against murine colitis.

Key points – the gut virome and mycobiome

- The virome and the mycobiome exhibit strong trans-kingdom interconnectivity with host immunity and with the bacteriome.
- Most gut viruses are phages; phageome diversity varies inversely with bacteriome diversity.
- Phage interactions with bacteria include not only the potential for bacterial lysis and viral replication but also the capacity to change the pathogenicity of commensal microbes.
- Phages may contribute directly and indirectly to host defense.
- Broad-spectrum antibiotics disrupt bacterial-fungal interactions and are a risk factor for candidemia in seriously ill patients.
- Fungal cell wall components are important drivers of mucosal cellular and humoral immunity.
- Clinically significant disturbances of the gut mycobiome occur in patients with IBD.

References

1. Shkoporov AN, Clooney AG, Sutton TDS, et al. The human gut virome is highly diverse, stable, and individual specific. *Cell Host Microbe*. 2019;26:527.
2. Shareefdeen H, Hill C. The gut virome in health and disease: new insights and associations. *Curr Opin Gastroenterol*. 2022;38:549–554.
3. Avellaneda-Franco L, Dahlman S, Barr JJ. The gut virome and the relevance of temperate phages in human health. *Front Cell Infect Microbiol*. 2023;13:1241058.
4. Harris HMB, Hill C. A place for viruses on the tree of life. *Front Microbiol*. 2021;11:604048.

5. Spencer L, Olawuni B, Singh P. Gut virome: role and distribution in health and gastrointestinal diseases. *Front Cell Infect Microbiol.* 2022;12:836706.
6. Liang G, Bushman FD. The human virome: assembly, composition and host interactions. *Nat Rev Microbiol.* 2021;19:514–527.
7. Cao Z, Sugimura N, Burgermeister E, Ebert MP, Zuo T, Lan P. The gut virome: a new microbiome component in health and disease. *EBioMedicine.* 2022;81:104113.
8. Mukhopadhya I, Segal JP, Carding SR, Hart AL, Hold GL. The gut virome: the 'missing link' between gut bacteria and host immunity? *Ther Adv Gastroenterol.* 2019;12:1–17.
9. Barr JJ, Auro R, Furlan M, et al. Bacteriophage adhering to mucus provide a non-host-derived immunity. *Proc Natl Acad Sci USA.* 2013;110: 10771–10776.
10. Gogokhia L, Buhrke K, Bell R, et al. Expansion of bacteriophages is linked to aggravated intestinal inflammation and colitis. *Cell Host Microbe.* 2019; 25:285–299.
11. Kernbauer E, Ding Y, Cadwell K. An enteric virus can replace the beneficial function of commensal bacteria.*Nature.* 2014;516:94–98.
12. Dallari S, Heaney T, Rosas-Villegas A, et al. Enteric viruses evoke broad host immune responses resembling those elicited by the bacterial microbiome. *Cell Host Microbe.* 2021;29:1014–1029.
13. Zhang F, Aschenbrenner D, Youn Yoo J, Zuo T. The gut mycobiome in health, disease, and clinical applications in association with the gut bacterial microbiome assembly. *Lancet Microbe.* 2022; 3:e969–e983.
14. Jiang TT, Shao T-Y, Ang WXG, et al. Commensal fungi replace the protective benefits of intestinal bacteria. *Cell Host Microbe.* 2017;22:809–816.

Gastroenterology

12 Therapeutic modification of the microbiome

HEALTHCARE

The Gut Microbiome

Diet is the most ancient means of modifying the microbiota with human breast milk being the first functional food consumed, providing beneficial effects on the physiology and microbiota of the neonate beyond its nutritional value. Likewise fermented foods, devised originally for food preservation, provide both nutrition and the functional benefits of microbial metabolites.[1] Here, we address non-antibiotic strategies adopted specifically to modify the microbiota for the prevention or treatment of disease. Figure 12.1 summarizes currently available therapeutic approaches varying in specificity from whole fecal microbiota transplantation (FMT) to the deployment of specific microbial metabolites and phages.

Fecal microbial transplantation

Although the first fecal microbiota product to be approved by the US Food and Drug Administration (FDA) was in 2022, FMT is not new. FMT is an example of science catching up with an old clinical observation. It was used in ancient times and was also known to be effective for treating pseudomembranous colitis even before the

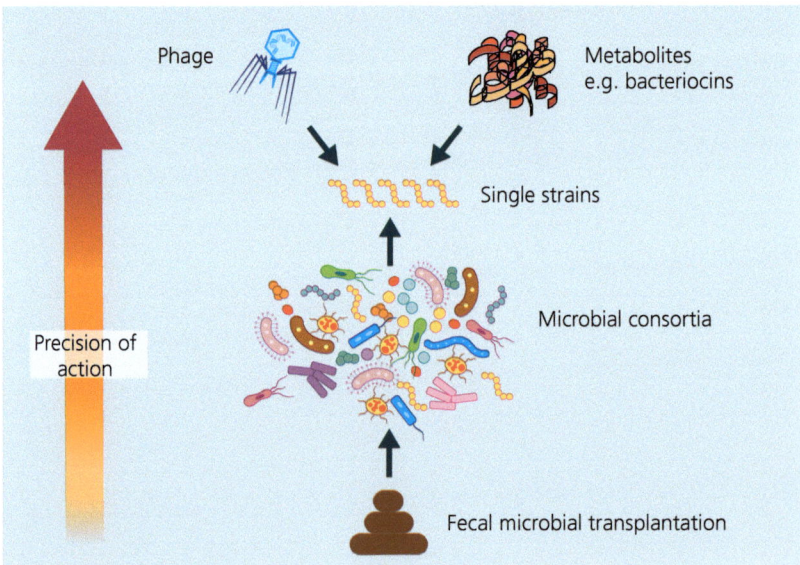

Figure 12.1 Schematic overview of potential non-dietary strategies for manipulating the gut microbiome.

causative agent (*Clostridioides difficile*) for that condition was identified (Figure 12.2). FMT is currently approved for the treatment of recurrent *C. difficile* infection only, for which it is remarkably efficacious.[2-7]

The mechanism of action of FMT in *C. difficile*-associated disease probably involves competitive exclusion of *C. difficile* by donor microbes leading to reduced toxin production, but other factors may include the restoration of protective microbes and modulation of the recipient's microbiome by phages and/or by donor microbes and metabolites.[8]

FMT has been tried in several other clinical settings but seldom with the same efficacy as that observed in *C. difficile*-associated disease.[8,9] However, encouraging results have been reported in patients undergoing treatment with immune checkpoint inhibitors for cancer, and in ulcerative colitis (particularly when administered repeatedly) but not in Crohn's disease.

Fecal transplant material is screened for known transmissible agents. Consequently, the risk of infectious diseases is low, although continual vigilance is essential as recently recognized by FDA safety alerts on the observed transmission of toxigenic *E. coli*.[10] In addition, the long-term risks of FMT are unknown. Since inflammatory, metabolic and behavioral phenotypes are transferable by FMT in experimental rodents, there is a theoretical potential for human-to-human transfer of the risk of various chronic diseases.

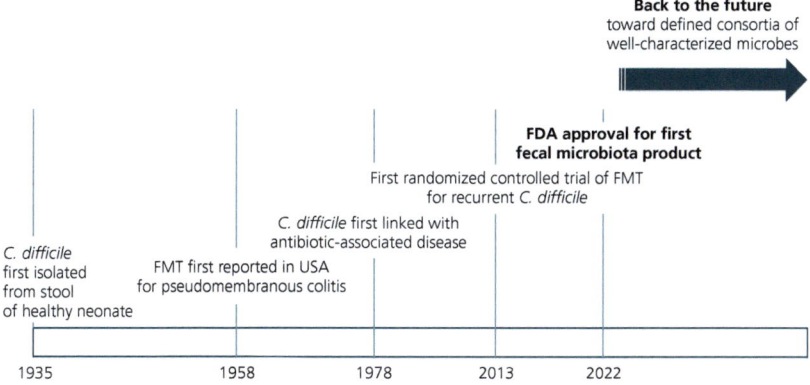

Figure 12.2 Historic context and timeline events prior to the approval of fecal microbial transplantation for recurrent *C. difficile* infection.[2-7]

For example, there is persuasive evidence that certain fecal microbiota may represent a risk factor for the development of colorectal cancer. It seems appropriate that recipients of FMT should be informed of this risk, albeit theoretical at present.

The future of microbial transplantation may be more intriguing than its past. Crude fecal preparations will probably soon be eclipsed by artificial stool comprised of fully characterized microbes. In the future, to ensure safety and to enhance efficacy, transplanted microbes will be carefully matched to each clinical indication and will be informed by analysis of the recipient's baseline microbiota.

Probiotics and live biotherapeutic products

Microbes – single or in combinations – have long been explored to enhance the gut microbiota for potential health benefits. Confusion has arisen because of the diversity of such organisms, the range of claims for potential benefits, the varying levels of scientific support and imprecise use of terminology.

The term probiotic may be used if three criteria are fulfilled: it must be live, consumed in adequate amounts, and confer a health benefit. However, these criteria are seldom adhered to.[11] Rarely have dose-response data been presented for such products and frequently the term probiotic has been used without any clinical evidence for a health benefit. In many instances, the claim for a probiotic is based only on in-vitro or preclinical data. Often the word probiotic has been misused without any evidence. In particular, food-grade or 'generally regarded as safe' microbes, such as bifidobacteria and lactobacilli, are not synonymous with probiotics. Moreover, since probiotics are highly strain dependent, and not restricted to bacteria (they may include yeasts such as *Sacharomyces boulardii*), meta-analyses of probiotic studies are confounded.

Other potentially confusing terms, such as 'next-generation probiotics' and pharmabiotics, have been used interchangeably with the term 'live biotherapeutics', but the regulatory landscape surrounding microbial therapeutics has recently changed, in part to address some of the confusing terminology.[12] Probiotics are now distinguished from live biotherapeutic products (LBPs) by regulatory agencies in the USA (FDA) and in Europe (Table 12.1).

TABLE 12.1

Regulatory overview of microbial therapeutic products

	Probiotics	**Live biotherapeutic products**
Definition	'Live microorganisms which when administered in adequate amounts confer a health benefit on the host' (FAO/WHO expert group)	Products containing live microorganisms that are 'applicable to the prevention, treatment, or cure of a disease or condition in human beings' (US FDA)
Regulatory category	Supplements & food for specific medical purposes	Drug
Target population	Healthy people	Disease

Notwithstanding the regulatory distinction between probiotics and LBPs, some probiotics have been shown in high-quality clinical trials to confer a benefit in certain diseases. The most noteworthy examples of convincing efficacy of some but not all putative probiotics have been reported in people with antibiotic-associated diarrheal disorders, necrotizing enterocolitis (NEC), and childhood diarrheal disease in developing countries. Some cases of irritable bowel syndrome and ulcerative colitis but not Crohn's disease may also benefit from specific microbial strains.

With clarification of regulatory terminology and corresponding guidelines, it is hoped that the future development of LBPs and preferably all forms of microbial therapeutics will adhere to rigorous traditional principles of therapeutics including a requirement for rigorously designed and controlled clinical trials to demonstrate benefit-risk ratio.

Genetically modified microorganisms

Genetically modified organisms (GMOs) have long been used to study biological processes. In the context of microbiome-based therapeutics, a GMO is a microbe that has been engineered to express a desired function, such as the production of a specific metabolite in vivo.[13-15] The range of possible applications is enormous. Thus, commensals or food-grade bacteria including probiotics have been used experimentally to correct host metabolic deficiencies and to deliver to the gut a diversity of bioactive molecules such as vaccines, enzymes, small molecule drugs and cytokines. A pioneering application was the deployment of *Lactococcus lactis* engineered to constitutively produce interleukin-10 (IL-10) in vivo in the gut to downregulate inflammation (colitis).

GMOs may also be designed to act as sensors, i.e. diagnostics to detect biomarkers of disease. Thus, microbes have evolved a multitude of genetically encoded sensors to monitor their environment, and these may be redirected for medical diagnosis. For example, the gut-adapted bacterium *E. coli Nissle* has been engineered to sense thiosulfate produced during inflammation. When the thiosulfate is sensed by the bacterium, the expression of a fluorescent protein is activated which may be detected in the feces. Oral administration of engineered bacteria has been used to sense various dietary constituents and microbial metabolites in the gut.[15] However, engineering either a drug delivery system or a sensory/diagnostic function is not limited to bacteria; yeast organisms have also been used.

Among the constraints on the technology is the imperative to ensure that the engineered organism is safe, that it cannot transfer genetic material to other environmental bacteria, and that it can be controlled if no longer needed or it escapes into the wider environment. Biocontainment strategies vary and include introducing a genomic mutation linked to a dependency of the GMO on a nutrient that is not accessible in the wider environment or equipping the GMO with a self-destruct mechanism if triggered by a specific nutrient. Another constraint on the technology is that GMOs cannot reliably sense disease biomarkers if they are not transported across the membrane or if they induce secondary chemical or physical changes inside the microbial cell. Other considerations in the design and choice of microbe include its adaptability to the target location within the gut and whether the genetic payload adversely affects its transit and survival in the gut.

Phage therapy

The growing global threat of antimicrobial resistance (AMR) is leading to a renaissance of interest in phages as alternatives to antibiotics. The appeal of phages as therapeutics is their bacterial specificity and consequently lack of collateral damage to the endogenous microbiome or to host cells. In some instances, phage may also penetrate bacterial biofilms.

Although phages have been used as therapeutics in former Soviet Union countries for about a century, their use in Western countries has been limited to research and/or on a compassionate use basis when other therapies have failed.[16,17] Several high-profile examples of the use of phage against multiple drug-resistant organisms such as *Acinetobacter baumannii*, *Mycobacterium abscessus*, and *Pseudomonas aeruginosa* causing lung infections in cystic fibrosis have highlighted the therapeutic use of phage. However, impressive anecdotal case reports are no substitute for clinical trials which are still lacking. Moreover, there are significant logistical hurdles to overcome before phage therapy can operate at a scale sufficient for widespread use.

Phages which must be grown in bacterial hosts may be difficult to produce at scale. More importantly, matching the correct phage or combination of phages for a particular bacterial target is time-consuming and labor-intensive. One approach is to produce in advance a cocktail of phages that are known to kill the most prevalent clinical strains of common pathogens. Phages are plentiful in the environment and are readily sourced in sewage and farmland. Large phage banks store phages against common pathogens. In contrast to antibiotics, the self-replicating property of phage implies that levels in the body may increase rather than decrease over time and confound dose calculations, creating uncertainty regarding long-term safety. To reduce the risk of resistance developing, a combination of phages is usually required.

Despite these hurdles, the threat of AMR is driving increasing interest in the use of phage along with technologically innovative approaches to optimize their efficacy, such as phage-mediated delivery of CRISPR-Cas to irreparably damage bacterial DNA.

Bacteriocins

Bacteriocins are naturally occurring small antibacterial peptides that are ribosomally produced by most bacteria, including commensals in the gut. A bacteriocin is not lethal to the producer microbe. Some bacteria can produce several types of bacteriocin. Bacteriocins protect a microbe's ecological niche, enabling the microbe to survive in a polymicrobial environment because they can eliminate competitors including colonizing pathogens. Thus, bacteriocins may be exploited as microbiome editing tools or as smart antimicrobials in the battle against AMR.[18]

Bacteriocins have been known for over a century. They have been deployed in the agri-food industry, particularly as products of lactic acid bacteria, and used for bio-preservation and fermentation processes. They are generally regarded as safe. However, the diversity of bacteria now known to exist in the gut means that a broad range of antimicrobial activity can be 'mined' from the gut microbiota. Some bacteriocins have an exquisitely narrow spectrum of activity which, in contrast to antibiotics, means that they could be selected to target a specific pathogen without causing collateral damage to the normal microbiota. Of course, bacteriocins are highly diverse in chemical structure and vary in their sensitivity to gastrointestinal proteolytic enzymes. Thus, there are numerous logistical and regulatory hurdles to be overcome before a bacteriocin can be developed as a drug for any clinical indication.

Key points – therapeutic modification of the microbiome

- The most accessible strategy for modifying the microbiome is by diet (human breast milk in infancy and later diversification with plants and fermented foods).
- FMT is an ancient strategy but only received regulatory approval recently for use in recurrent *C. difficile* infection.
- Potential hazards of FMT are not restricted to infectious diseases; the risk of transmission of inflammatory, metabolic and behavioral phenotypes is suggested by animal studies.
- Crude fecal transplants are likely to be eclipsed by use of consortia of fully characterized microbes.
- Use of single or combinations of live microbes to enhance the gut microbiota for health benefits has been confounded by varying levels of scientific support and imprecise use of terminology.
- Probiotics are now distinguished from live biotherapeutic products (LBPs) for regulatory purposes in the USA and Europe.
- Increasingly interest is becoming focused on GMOs, bacterial metabolites including bacteriocins and phage therapy.

References

1. Marco ML, Sanders ME, Gänzle M, et al. The International Scientific Association for Probiotics and Prebiotics (ISAPP) consensus statement on fermented foods. *Nat Rev Gastroenterol Hepatol.* 2021;18:196–208.
2. Hall I, O'Toole E. Intestinal flora in newborn infants with a description of a new pathogenic anaerobe, Bacillus difficilis. *Am J Dis Child.* 1935;49:390.
3. Bartlett JG, Moon N, Chang TW, et al. Role of *Clostridium difficile* in antibiotic-associated pseudomembranous colitis. *Gastroenterology.* 1978;75:778–782.

4. George WL, Sutter VL, Goldstein EJ, et al. Aetiology of antimicrobial-agent associated colitis. *Lancet.* 1978;1:802–803.
5. Eiseman B, Silen W, Bascom GS, Kauvar AJ. Fecal enema as an adjunct in the treatment of pseudomembranous enterocolitis. *Surgery.* 1958;44:854–859.
6. Van Nood E, Vrieze A, Nieuwdorp M, et al. Duodenal infusion of donor feces for recurrent *Clostridium difficile*. *N Engl J Med.* 2013;368:407–415.
7. Bartlett JG. *Clostridium difficile*: history of its role as an enteric pathogen and the current state of knowledge about the organism. *Clin Infect Dis.* 1994;18(suppl 4):S265–S272.
8. Murphy CL, Zulquernain SA, Shanahan F. Faecal microbiota transplantation (FMT) - classical bedside-to-bench clinical research. *QJM.* 2023;116:641–643.
9. Marrs T, Walter J. Pros and cons: is faecal microbiota transplantation a safe and efficient treatment option for gut dysbiosis. *Allergy.* 2021;76:2312–2317.
10. United States Food and Drug Adminstration: Fecal microbiota for transplantation: safety alert – risk of serious adverse events likely due to transmission of pathogenic organisms. FDA; 2020. fda.gov/safety/medical-product-safetyinformation/fecal-microbiotatransplantation-safety-alertrisk-serious-adverse-eventslikely-due-transmission, last accessed March 5 2024.
11. Shanahan F, Hill C. Language, numeracy and logic in microbiome science. *Nat Rev Gastroenterol Hepatol.* 2019;16:387–388.
12. Cordaillat-Simmons M, Rouanet A, Pot B. Live biotherapeutic products: the importance of a defined regulatory framework. *Exp Mol Med.* 2020;52:1397–1406.
13. Ciocan D, Elinav E. Engineering bacteria to modulate host metabolism. *Acta Physiol (Oxf).* 2023;238(3):e14001.
14. Basarkar V, Govardhane S, Shende P. Multifaceted applications of genetically modified micro-organisms: a biotechnological revolution. *Curr Pharm Des.* 2022;28:1833–1842.
15. Landry BP, Tabor JJ. Engineering diagnostic and therapeutic gut bacteria. *Microbiol Spectrum.* 2017;5(5):10.1128/microbiolspec.BAD-0020-2017.
16. Summers WC. The strange history of phage therapy. *Bacteriophage.* 2012;2:130–133.
17. Law, N, Aslam S. Phage therapy: primer and role in the treatment of MDROs. *Curr Infect Dis* Rep. 2020;22:31.
18. Simons A, Alhanout K, Duval RE. Bacteriocins, antimicrobial peptides from bacterial origin: overview of their biology and their impact against multidrug-resistant bacteria. *Microorganisms* 2020; 8:639.

Index

adipose tissue 53
age
  changes in microbiome 17, 19, 42–8, 124
  colonization of GF animals 27
alcohol use 102
5-aminosalicylic acid (5-ASA) 100
androgens 34
animal models
  germ-free 24–8
  IBD 78
antimicrobial resistance (AMR) 108–15, 135
antimicrobials
  antibiotics 13, 24, 58, 70, 90, 108, 123
  bacteriocins 15, 136
  phages 135
artificial sweeteners 60
assembly *see* colonization
autism 35, 71
autoimmune disease 34, 76, 83
autonomic nervous system 68

bacteriocins 15, 136
bacteriophages 81, 118–22, 126
  therapeutic 135
Bacteroides 43, 88
Bifidobacterium 43, 45, 94, 103
bile acids 17, 81, 91
birth 42–3, 123
body-site specificity 14–15
bowel disease *see* intestinal disease
brain
  brain-gut signaling 68–72
  disorders 35, 71, 99, 100–1
  in GF animals 25
breast cancer 91
breastfeeding 44, 123–4

cancer 88–95
  causation 34, 88, 90–1, 132
  host defense 89, 90, 92, 93–4
  intratumoral bacteria 89–90, 92
  management 89, 90, 93–4
Candida 123, 124, 125
Caudovirales 81, 118
celiac disease 83
cesarean section 43
characteristics 12–20
chemotherapy 89, 90, 93
chronic intestinal diseases 77–84, 125
chronic kidney disease 36, 62
chronic NCCDs generally 76–7
cigarettes 78, 102–3
circadian rhythm 14
Clostridioides difficile 60, 130–1
colon/rectum
  C. difficile 60, 130–1
  cancer 88, 91, 93, 132
  IBD 77–81, 125, 131
  IBS 71, 81–3, 133
  microbiome 14, 15
colonization 13, 17, 42–3, 76, 118, 123
  in GF animals 27

communication about AMR 112
cooking 59
Crohn's disease 77–81, 125

Debaryomyces hansenii 125
diagnostic uses 93, 134
diarrhea 102, 133
diet and nutrition 52–63, 130
  and cancer 94
  fat metabolism 52–3, 56
  fiber/SCFA 25, 52, 53, 54–5
  and fungi 123, 124
  gas production 53–4, 57
  GF animals 25
  infants 43–5, 123–4
  malnutrition 54, 58
  microbial metabolism 56, 70–1
  obesity 58
  older people 48
  processed food 58–9, 60
  specialized diets 48, 59, 61–2, 94
digoxin 100
disease
  and aging 19, 47–8
  cancer 34, 88–95, 132
  chronic NCCDs 76–7
  and diet 54, 56
  and fungi 124, 125
  GF animal models 27
  gut-brain axis 71
  intestinal 71, 77–84, 125, 130–1, 133
  neurodegenerative 35, 71, 99, 100–1
  and phages 122
  variance in microbiome 34–6

diversity 14, 16, 45, 47, 61, 76, 79, 118, 124
L-Dopa 99, 100–1
drug discovery 103
drug interactions 98–104
 chemotherapeutics 89, 93
drug resistance (AMR) 108–15, 135

elderly people 19, 46–8, 124
emulsifiers 60
enterochromaffin cells 54, 71
Escherichia coli 88, 131
Escherichia coli Nissle 134
estrogen 34
evolution 13

farming 108
fat metabolism 52–3, 56
fecal microbial transplants (FMT) 58, 71, 93–4, 130–2, 137
fermented foods 61
fetal immune system 42
fiber (and SCFA) 25, 52, 53, 54–5, 70
food additives 59, 60
fungi 122–5, 126
Fusobacterium nucleatum 88, 89

gas production 53–4, 57
gender 34
genetically modified organisms 133–4
genetics
 horizontal gene transfer 112, 121
 IBD 78, 125
germ-free (GF) animals 24–8

gluten intolerance 83

Helicobacter pylori 81, 88, 124
hepatic cancer 91
hepatic encephalopathy 35, 70
histidine 56
horizontal gene transfer 112, 121
horizontal transmission 13, 33
host-microbe signaling 15–17, 54–5, 70–1, 120–2
human milk oligosaccharides 43–4
hygiene hypothesis 76

immune system 15–16
 autoimmunity 34, 76
 and cancer 89, 90, 93–4
 fetal 42
 and fungi 123, 124–5
 neonatal 76
 and phages 120–2
immunotherapy 89, 93–4
individuality 13
industrialized vs non-industrialized societies 36–8, 76–7
infants 13, 17, 32, 42–5, 76, 123–4
inflammatory bowel disease (IBD) 77–81, 125, 131
interferon 122
intestinal disease
 C. difficile infection 60, 130–1
 cancer 88, 91, 93, 132
 celiac 83
 IBD 77–81, 125, 131
 IBS 71, 81–3, 133
irinotecan 93

Irish Travellers 37–8
irritable bowel syndrome (IBS) 71, 81–3, 133

kidney disorders 36, 56, 62

Lactobacillus reuteri 94
Lactococcus lactis 134
lactulose 102
laxatives 100, 102
lifestyle 14, 33, 37–8, 76–7
lipopolysaccharide (LPS) 56, 71
live biotherapeutic products 132–3
liver disease 35, 70, 91
low FODMAP diet 61
low protein diet 62

Malassezia 123, 125
malnutrition 54, 58
maternal transmission 42–3, 123
Mediterranean diet 48, 61, 94
melanoma 93, 94
men 34
metabolism
 of fat 52–3, 56
 in IBD 81
 microbial 15, 17, 25, 56, 70–1
 of sex hormones 34
 see also drug interactions
metformin 35, 100
microbe-microbe signaling 15
microbial depletion technique 24
Microviridae 81, 118
migration studies 76–7
mycobiome 122–5, 126

Index

necrotizing enterocolitis 45
neonates 13, 32, 42–3, 44–5, 76, 123
neurodegenerative diseases 35, 71, 99, 100–1
neurotransmitters 70–1
normal microbiomes 32–8
number of microbes 12, 118, 123
nutrition *see* diet and nutrition

obesity 58
older people 19, 46–8, 124
'One Health' approach to AMR 114
oxalate 56

pain in IBS 71
pancreatic cancer 90, 93
Parkinson's disease 71, 99, 100–1
phages 81, 118–22, 126
  therapeutic 135
plasticity 13
pregnancy 43
prenatal period 42
preterm infants 32, 44–5
probiotics 70, 94, 132–3, 137
processed foods 58–9, 60

proton pump inhibitors (PPIs) 100, 102
pseudomembranous colitis 60, 130–1

renal disorders 36, 56, 62
restrictive diets 61

Saccharomycetales 123–4
serotonin 70–1
sex 34
short-chain fatty acids (SCFA) 25, 52, 54–5, 70, 81
signaling
  brain-gut 68–72
  host-microbe 15–17, 54–5, 70–1
  host-phage 120–2
  microbe-microbe 15
site specificity 14–15
small intestinal bacterial overgrowth (SIBO) 82–3
small intestine 14, 15, 91
  celiac disease 83
smoking 78, 102–3
socioeconomic sources of variance 33–4, 36–8, 47, 76–7
spatial arrangement 14–15
stomach 15
  cancer 88, 91
sulfasalazine 100

temporal changes 17, 19, 42–8, 124
tobacco 78, 102–3

transfer of microbiota
  human to GF animals 27
  to neonates 13, 42–3, 44, 123
trehalose 60
trimethylamine (TMA) 56
tryptophan 56, 70–1, 81, 94

ulcerative colitis 77–81, 125, 131
uniqueness 13

vaginal delivery 43, 123
vagus nerve 68
variance in microbiota and disease 34–6
  preterm infants 44–5
  and signaling 16
  sources 14, 32–4, 36–8, 47, 76–7
vertical transmission 42–3, 123
very low calorie diet 61
viruses
  and cancer 88
  gut virome 81, 118–22, 126
  therapeutic 135

weight changes 53, 58, 103
women
  disease 34, 91
  maternal transmission 42–3, 123

xantham gum 60